WORLD BANK DISCUSSION PAPER NO. 387

The Agrarian Economies of Central and Eastern Europe and the Commonwealth of Independent States

Situation and Perspectives, 1997

Csaba Csaki
John Nash

The World Bank
Washington, D.C.

Discussion Papers present results of country analysis or research that are circulated to encourage discussion and comment within the development community. The typescript of this paper therefore has not been prepared in accordance with the procedures appropriate to formal printed texts, and the World Bank accepts no responsibility for errors. Some sources cited in this paper may be informal documents that are not readily available.

Cover photo: Combine harvesting wheat by Yosef Hadar/World Bank, 1983.

ISSN: 0259-210X

Csaba Csaki is lead agricultural adviser and John Nash is principal economist in the Rural Development and Environment Sector Unit of the World Bank's Europe and Central Asia Department.

Library of Congress Cataloging-in-Publication Data

Csaki, Csaba.
    The agrarian economies of Central and Eastern Europe and the
 Commonwealth of Independent States : situation and perspectives, 1997
 / Csaba Csaki, John Nash.
        p.   cm. — (World Bank discussion paper, ISSN 0259-210X ; no.
 387)
    Includes bibliographical references (p.   ).
    ISBN 0-8213-4238-X
    1. Land reform—Europe, Eastern—Longitudinal studies.   2. Land
 reform—Former Soviet republics—Longitudinal studies.
 3. Agriculture—Economic aspects—Europe, Eastern—Longitudinal
 studies.   4. Agriculture—Economic aspects—Former Soviet republics—
 Longitudinal studies.   I. Nash, John D., 1953–   . II. Title.
 III. Series: World Bank discussion papers ; 387.
 HD1333/E852C77   1998
 333.3'147—dc21   98-24255
                                                           CIP

Contents

Foreword

The agrarian economies of Central and Eastern Europe and the former Soviet Union are, at present, undergoing a systemic change and transformation. It is becoming increasingly apparent that the countries concerned have made the right choice in setting their overall goals and the direction of the transition. Under the present economic and political conditions in the region there is no alternative to the creation of a market economy based on private ownership. However, given the developments of the past five years, it is clear that the transformation is a far more complicated and complex process than anyone imagined in 1990-91. Seven years after launching the reform processes it still cannot be said that the transformation is complete. The region's agrarian economy has not yet pulled out of one of the most difficult stages in its history.

This paper provides benchmark information for studying problems of this difficult transition in Central and Eastern Europe and for the development of specific agricultural projects in the individual countries of the region. This study is also an important foundation for developing World Bank assistance strategies in the rural sector for the region as a whole, as well as for the individual countries. While this study was developed for World Bank use, I believe it will be useful to the international donor community and others active in studying the transition process in the rural sector in Eastern Europe and Central Asia.

The volume presents an overview of the agricultural situation and the status of the reforms in the agrarian sector in Central and Eastern Europe and the countries of the Commonwealth of Independent States (CIS). This assessment is based on the FAO agricultural database as well as country-specific analysis prepared by the relevant World Bank staff members. To evaluate of the status of reforms, a special methodology has been developed at the World Bank to compare agricultural reform performance across all countries in Eastern Europe and Central Asia. The general overview and the country-specific assessment presented in the form of policy matrices is supplemented with an aggregated statistical database.

March 16, 1998

Kevin Cleaver
Sector Director, ECSRE

Abstract

The agrarian economies of Central Eastern Europe and the former Soviet Union are undergoing a systemic change and transformation. Looking back it can be seen that the countries concerned made the right choice in setting their overall goals and policies for transition to a market economy. Under the present economic and political conditions in the region there is no alternative to the creation of a market economy based on private ownership. However, given the developments of the past seven years, it is clear that the initial expectations for transformation were overly optimistic and the transition process is far more complicated and complex than anyone imagined in 1990. The region's agrarian economy is still struggling to adjust to economic reality.

The region has a substantial part of the world's agricultural resources. In 1994 the countries concerned comprised 13% of the world's area suitable for agricultural production and 19.7% of arable land. Between 1990 and 1995, the share of GDP produced in agriculture at the regional level fell by an annual average of 5.9% and during this time it was around 60% below the level of 1989-90. The region makes a substantial contribution to world output in practically all of the main agricultural products. Yet despite abundant natural resources, the region still plays a small role in the world food trade. Overall the region continues to be a net importer of agricultural products. The negative balance of the Commonwealth of Independent States (CIS) countries has been reduced slightly, while in the case of Central and Eastern Europe (CEE) it has grown. Perhaps the most significant structural change is that the CIS countries, and Russia in particular, have become one of the world's biggest meat importing regions. In place of the massive grain imports characteristic of the Soviet period, Russia now mainly buys meat.

In the CEE countries, the process of sector transformation will probably be completed within the next four or five years. Together with the anticipated acceleration of general economic development, this could lead to the stabilization of food production, more efficient production, and a greater upswing driven by potential comparative advantages. It is far more difficult to predict changes in the CIS countries. It seems probable that further difficult years lie ahead for the sector in this region, compounded by the struggle between conservative and progressive forces. The reforms in many of these countries will probably continue to advance only slowly.

Considering the overall prospects for the region's trade performance, our conclusion is that, in the region as whole, net imports will fall slightly if even moderate general growth of agriculture is realized, and the region will continue to be a net food importer. It would appear that fuller exploitation of the very considerable natural resources must remain for the more distant future. It is likely that, because of differences in the recovery rate of production levels among the individual countries and the pace of reforms, differences within the region will grow.

Preface

This volume is a compilation of a year's work analyzing the problems of the rural sector at the regional level. The study is focused on agricultural production and trade, and related policies. Although we recognize the importance of the social and natural resource management aspects of the agricultural transition, these issues are not addressed in this report. The work presented in this paper was managed and coordinated by Csaba Csaki, who is the author of the overview as well as the creator of the methodology used to compare agricultural reform performance in the individual countries. He was assisted by John Nash in compiling the country-specific policy matrices into a consistent framework. Country-specific matrices were prepared for all the countries of the region except Cyprus, Portugal, Turkey, the Czech Republic, and the Federal Republic of Yugoslavia.

The individual country policy matrices were prepared by the following Task Managers: Albania (S. Kodderitzsch), Armenia (C.Csaki, M.Lundell), Azerbaijan (R. Southworth), Belarus (C. Csaki), Bosnia and Herzegovina (M. Koch), Bulgaria (J. Nash), Croatia (S. Kodderitzsch), Estonia (C. Csaki), Georgia (C.Csaki, I.Shuker), Hungary (C.Csaki), Kazakhstan (R. Southworth), Kyrgyz Republic (M. Mudahar), Latvia (C.Csaki), Lithuania (C.Csaki), FYR Macedonia (M. Nightingale), Moldova (C.Csaki), Poland (R. Lacroix), Romania (H. Gordon), Russia (K.Brooks), Slovak Republic (M. Hermann), Slovenia (M. Nightingale), Tajikistan (T. Sampath), Turkmenistan (K.Brooks), Ukraine (C.Csaki, M.Lundell), and Uzbekistan (M. Nightingale).

Kevin Cleaver and Laura Tuck provided valuable comments and suggestions throughout the study. The guidance of Geoffrey Fox in the development of the first version of the methodology used in the country comparisions is very much appreciated. The preparation of the overview, as well as the statistical database was supported by Achim Fock. Alan Zuschlag provided editorial assistance.

Executive Summary

This report seeks to provide a brief overview of agricultural economies in the region. It identifies where the agrarian economies of Eastern Europe and Central Asia stand today, the direction in which they are heading, the rate at which production can be expected to recover, and how this influences the region's behavior on international agricultural markets. Although we recognize the importance of the social and natural resource management aspects of the agricultural transition, these issues are not addressed in this report.

The region has a substantial part of the world's agricultural resources[1]. In 1994 the countries concerned comprised 13% of the world's area suitable for agricultural production and 19.7% of arable land. The region makes a substantial contribution to world output in practically all of the main agricultural products. Yet despite abundant natural resources, the region still plays a small role in the world food trade. It is quite clear, in examining the possibility for increasing food production from the global angle, that the region, where around 20% of the potential global resources are located, holds significant promise for meeting the substantial new food demands forecasted for the first half of the next century. Furthermore, this can be achieved without the risk of causing serious harm to the natural environment.

Between 1990 and 1995, the share of GDP produced in agriculture at the regional level fell by an annual average of 5.9% and during this time it was around 60% below the level of 1989-90. Despite the decline in production, the region's share of world trade did not shrink substantially and, in the case of some products, there was even an unquestionable increase. This was made possible (or required) by the fall in domestic consumption and by the new situation created with the disintegration of the Soviet Union.

While the structure of the region's agricultural exports and imports has changed considerably as a result of declining consumption, the balance of agricultural trade for the region as a whole did not deteriorate. Overall the region continues to be a net importer of agricultural products. The negative balance of the Commonwealth of Independent States (CIS) countries has been reduced slightly, while in the case of Central and Eastern Europe (CEE) it has grown. Perhaps the most significant structural change is that the CIS countries, and Russia in particular, have become one of the world's biggest meat importing regions. In place of the massive grain imports characteristic of the Soviet period, Russia now mainly buys meat. This is quite clearly a more favorable solution from an economic viewpoint since the large quantity of grain purchased in earlier decades by the Soviet Union was used in animal husbandry with very low efficiency. At the same time, the CIS countries are increasingly appearing on world markets as grain exporters, beyond the confines of their former trading patterns with each other in the Soviet Union.

In 1990-91, the region set out on the path of creating market economies based on private property. There has been a great deal of similarity between countries in terms of the tasks

---

[1] For purposes of this paper "the Region" comprises all the former socialist economies of Central and Eastern Europe and Central Asia. For a fuller description of the countries and groupings listed in this report, please refer to **Box 1** in the main text.

required to achieve such an outcome. However, there are quite big differences when it comes to the pace of realization and the manner of implementation. Our analysis of the region of as a whole indicates[2]:

- The reform process is considerably more complicated and complex than originally expected and results of the reform process to date have only achieved a part of those original expectations. A relatively rapid growth in production, similar to the results of the Chinese reforms, has not occurred.

- The pace of transformation of the agrarian sector and the rural economy is lagging behind the rate of changes in the economy as a whole.

- The biggest transformation has taken place in the price and market environment, while there is a substantial lag in solving the financing problems of agriculture and in the area of institutional reforms. The rapid spread of private, family farms has still to come; to a large extent the inherited large-unit structures for agricultural production have survived the changes.

- There are considerable differences in the reform performances of the individual countries. As a group, the CEE countries are quite clearly far more advanced than the CIS countries.

- The transformation of agriculture is most advanced in CEE countries and, in particular, in the EU candidate countries. Even here, reform of the agrarian sector has not yet been fully accomplished. The results of our analyses agree with the EU evaluation in finding that further reforms are needed, principally in the institutional systems and in rural financial markets, but also in land reform and the transformation of the inherited economic structures. This process is still unfinished in practically all of the countries destined for EU accession in the next five years.

- The transformation of agriculture in the CIS countries is still in its early stages. Solving the first tasks for agrarian reforms is still the first item on the agenda here. The system of institutions and instruments of the planned economy has not yet been fully dismantled, and functioning agrarian markets have still not come into being.

Many valuable conclusions can be drawn from the analysis of the experiences of the countries leading in the transformation, including:

- A general economic upswing can greatly assist the agricultural reforms. The greatest progress has been made in transforming the sector in those countries where a comprehensive economic recovery is also underway.

---

[2] The description of the status of reforms for each country matrix was compiled by the World Bank staff most familiar with that particular country's agricultural policies. Numerical ratings were then assigned to each of the five reform categories in accordance with the criteria listed in **Table 5**. These ratings were then revised in several review sessions to improve consistency of rankings.

- Development in the non-agricultural segment of the rural economy is of key importance for the recovery of agriculture. In the great majority of the countries that are the most advanced in the reforms, it has been the upswing of the rural economy surrounding agriculture that has made possible a substantial reduction in the numbers of people employed in agriculture, and allowed for improvements in the efficiency and competitiveness of agriculture itself.

- An important factor in the successes achieved in reforms is consistency, and the combined implementation of parallel steps in areas related to the reforms.

- The greatest reform progress has been made by those countries that have elected to quickly make wide and comprehensive reforms, despite the fact that these efforts are causing greater difficulties in the short-term. In most cases the appeals for a gradual approach appear to be a sign of the lack of will; this is especially the case in the CIS countries.

In all countries the process of agricultural reforms has been strongly influenced by day-to-day politics. Very often, politics still determine the pace and extent of reforms, at the expense of economic rationality. In general, there is a lack of a carefully considered, long-term strategies, and an objective and realistic evaluation of the economic consequences of the different possible solutions. As a result, the negative economic consequences of the transformation are greater than the unavoidable minimum, even in the most advanced countries.

Three factors will directly determine the course of agricultural production in this region, namely: a) the investments and current assets available; b) the rate of technological renewal, together with the pace of introduction of new advances in biology and higher-yield crop and livestock strains; and c) changes in the organization and the human environment of production. Yet progress in these three areas will, in turn, depend on wider economic and policy reforms, the results achieved in the transformation of agriculture, and the internal demand for agricultural produce.

In the CEE countries, the process of sector transformation will probably be completed within the next four or five years. Together with the anticipated acceleration of general economic development, this could lead to the stabilization of food production, more efficient production, and a greater upswing driven by potential comparative advantages. Although the effect of EU accession cannot be clearly quantified, it will probably give further stimulus for the growth of production in the countries concerned.

It is far more difficult to predict changes in the CIS countries. It seems probable that further difficult years lie ahead for the sector in this region, compounded by the struggle between conservative and progressive forces. The reforms in many of these countries will probably continue to advance only slowly. However, it is not likely that the decline in production will continue. The results of 1995-96 and the estimates for 1997 indicate that the region's agricultural output has already reached its lowest point. In the short-term, the most likely course is moderate growth alternating with stagnation. Substantial growth in production can only be expected as substantive progress is made in the reforms that directly determine the factors of production. This will probably happen only in the medium-term.

Considering the overall prospects for the region's trade performance, our conclusion is that, in the region as whole, net imports will fall slightly if even moderate general growth of agriculture is realized, and the region will continue to be a net food importer. It would appear that fuller exploitation of the very considerable natural resources must remain for the more distant future. It is likely that, because of differences in the recovery rate of production levels among the individual countries and the pace of reforms, differences within the region will grow.

A. OVERALL ANALYSIS

The agrarian economies of Central and Eastern Europe (CEE) and the former Soviet Union are undergoing a systemic change and transformation. Looking back it can be seen that the countries concerned made the right choice in setting their overall goals and policies for transition to a market economy. Under the present economic and political conditions in the region there is no alternative to the creation of a market economy based on private ownership. However, given the developments of the past seven years, it is clear that the initial expectations for transformation were overly optimistic and the transition process is far more complicated and complex than anyone imagined in 1990-91 (Csaki and Lerman, 1997). The region's agrarian economy is still struggling to adjust to economic reality.

While there is no "blueprint" for adjusting agricultural policies to the market economy, it is possible to determine certain salient features and strategic behavior that are conducive to effective transitions based on the experiences of countries further along in the transition process. This report seeks to provide a brief overview of agricultural economies in the region. It identifies where the agrarian economies of Eastern Europe and Central Asia stand today, the direction in which they are heading, the rate at which production can be expected to recover, and how this influences the region's behaviour on international agricultural markets. The analyses are based on statistics processed by FAO for each of the individual countries, and also on the results of analytical work being done with the support of the World Bank. It needs to be acknowledged, however, that there are questions of reliability and consistency between the statistical information from all countries mentioned in the report. The textual analysis of the situation and forecasts largely follow the two main groupings of the region: Central-Eastern Europe and the Commonwealth of Independent States. However, most of the tables given in the study also present the results of analysis with the two main country groups broken down into further subregions (see **Box 1**).

Box 1: Country Groupings Used in the Analysis:

The "region:" all former socialist countries of Europe, together with the former Soviet Union;

Central-Eastern Europe (CEE): the former socialist countries of Central-Eastern Europe, including the Baltic republics of the former Soviet Union, Albania, the member republics of ex-Yugoslavia and Albania;

 CEFTA: the five CEFTA member countries in 1996 (Poland, Hungary, Czech Republic, Slovak Republic; and Slovenia);

 EU-A5: the rest of the countries in the EU-accession process - Estonia, Latvia, Lithuania, Romania, and Bulgaria;

 Other Central-Eastern Europe: the remaining countries of Central-Eastern Europe;

CIS: all CIS member countries of the former Soviet Union;

 Russia

 European CIS: Belarus, Moldova, and Ukraine;

 Caucasus: Georgia, Armenia, Azerbaijan;

 Central Asia: Tajikistan, Turkmenistan, Kazakhstan, the Kyrgyz Republic, and Uzbekistan.

The Significance of the Agrarian Economy and Natural Resources in The Region

The agrarian economy has traditionally played an important role in the countries of this region (see **Tables 1** and **2**).

Table 1: Population, Agricultural Area, and Arable Land of Transition Countries
(Absolute and in Comparison to World and Total Transition Countries)

	Population, 1996			Agricultural Area, 1994			Arable Land, 1994		
	in mil	% of World	% of TC	in mil ha	% of World	% of TC	in mil ha	% of World	% of TC
TOTAL	414	7.3	100.0	649	13.3	100.0	267	19.7	100.0
Total CEE	129	2.3	31.1	74	1.5	11.4	50	3.7	18.6
Total CIS	286	5.0	68.9	575	11.8	88.6	217	16.0	81.4
CEFTA	66	1.2	16.0	32	0.7	4.9	24	1.8	9.0
EU-A5	39	0.7	9.4	28	0.6	4.3	19	1.4	7.2
Other CEE	24	0.4	5.8	13	0.3	2.0	7	0.5	2.5
Russia	148	2.6	35.8	220	4.5	33.9	130	9.6	48.9
European CIS	66	1.2	16.1	54	1.1	8.3	41	3.0	15.4
Caucasus	17	0.3	4.0	8	0.2	1.2	3	0.2	1.1
Central Asia	54	1.0	13.0	293	6.0	45.1	43	3.1	16.0

Source: FAOStat

Table 2: Role of Agriculture

	Ag. GDP as % of total GDP, 1995**	Agricultural Population	
		in Millions, 1996	in % of total Population, 1996
TOTAL*	10.8	74.0	17.8
Total CEE	9.2	23.1	17.9
Total CIS	11.8	50.8	17.8
CEFTA	6.7	11.9	18.0
EU-A5	15.8	5.7	14.8
Other CEE	17.4	5.5	22.8
Russia	9.2	18.0	12.1
European CIS	13.3	12.4	18.7
Caucasus	36.7	4.2	25.0
Central Asia	22.7	16.3	29.9

*Azerbajian, Croatia, Lithuania, FYR Macedonia, Moldova, Czech Rep., Yugoslav FR, Slovenia, Tajikistan, Turkmenistan not included
** Moldova is not available and not included
Source: WDR, WDI, FAOStat

- The share of agriculture in the employment and national income in these countries is far greater than the average for the developed countries. In 1995 the agrarian sector contributed 10.8 % of GDP for the region as a whole, while the proportion of the agricultural population was 17.8 % (**Figure 1**).

- However, there are very substantial deviations from country to country in regard to this larger sectoral role than is generally found in the developed countries. The share of agriculture in GDP is only 6.7 % in the CEFTA countries, compared to 36.7 % in the Caucasus, and 22.7 % in Central Asia.

The contribution of agriculture to GDP is much lower than the role it plays in employment. This is an indication of the low productivity of agriculture and efficiency below the international average.

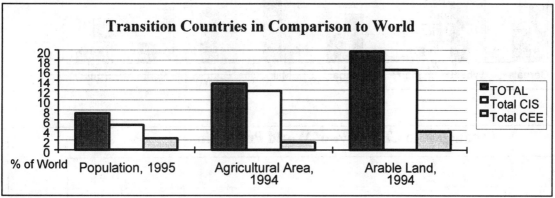

Figure 1

The East European region has a substantial part of the world's agricultural resources. In 1994 the countries concerned comprised 13% of the world's area suitable for agricultural production and 19.7% of arable land. The region makes a substantial contribution to world output in practically all of the main agricultural products (**Figures 2, 3**, and **4**). For the majority of products, this contribution is around 10-15%.

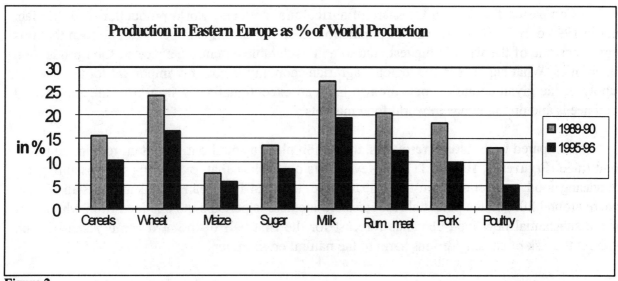

Figure 2

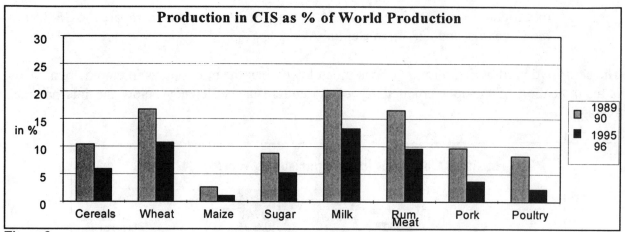

Figure 3

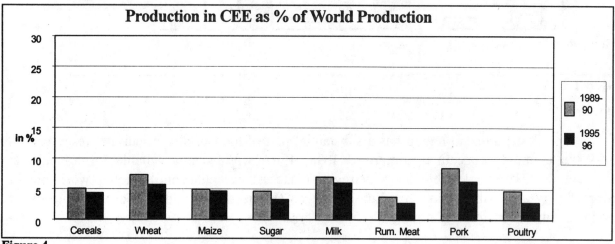

Figure 4

Comparing the data on the share of agricultural area and world production with the fact that in 1994 only 7.3% of the world's population lived in Eastern Europe, it can be seen that this region has one of the world's biggest, and to a considerable extent, unexploited food production capabilities. What happens in the region's agrarian economy is not only important for the internal supply of the given countries, but can also have an exceptionally big impact on meeting world food needs and on the trends in world food markets.

Compared to its natural resources, the region plays a small but important role in the world food trade (**Figures 9, 10,** and **11**). However, it is quite clear that, examining the possibility for increasing food production from the global angle, the East European and Central Asian region, where around 20% of the potential resources are located, is perhaps the one area from which part of the substantial new food demands forecast for the first half of the next century can be met, without the risk of causing serious harm to the natural environment.

The Food Production Situation in the Region

The agrarian economy of the region has been characterized in the 1990s by a considerable fall in production.

- Between 1990 and 1995 the share of GDP produced in agriculture at the regional level fell by an annual average of 5.9% and for the period as a whole was around 40% below the level of 1989-90 (**Table 3**).

- The fall in production affected essentially all countries, and all subregions of the region. The annual average decline was the smallest in Central-Eastern Europe (1.9% a year) and the biggest in the European CIS countries and the Caucasus (10% and 17.3% respectively) (**Figures 5** and **6**).

- Overall, in practically all countries the fall in agricultural output was less than the decline in the economy as a whole. As a result, the contribution of agricultural output to GDP grew throughout the region.

- The extent of the fall in production differed considerably, not only from country to country, but also from one product to another. In the CIS countries, in 1995-96 the output for all main agricultural products was below 1989-90 levels. The biggest fall can be found in animal products (30-60%). In Central-Eastern Europe the output of main products, including animal products, fell to a lesser extent. By 1995-96 the production of oil plants and vegetables exceeded 1989-90 levels (**Figures 7** and **8**).

There was a tangible decline in the contribution of the region and all its subregions to world output in its main agricultural products. In 1995-96 this rate at the regional level was around 30% lower than in 1989-90 and in some cases was as much as 50% lower (**Figures 2, 3**, and **4**).

Table 3: GDP Development

	Average Annual GDP Growth Rate			Average Annual Change in Agricultural GDP 1990-1995*
	1995*	1990-1995*	1997 forecast** (EBRD)	
TOTAL	-1.5	-7.0	2.0	-5.9
Total CEE	5.2	-0.7	3.5	-1.9
Total CIS	-5.8	-10.6	1.0	-7.5
CEFTA	5.6	0.3	4.5	-3.0
EU-A5	5.1	-4.5	-1.1	-1.7
Other CEE	1.9	n.a.	5.5	7.6
Russia	-4.0	-9.8	1.5	-6.3
European CIS	-12.0	-13.3	-1.5	-10.0
Caucasus	-7.2	-22.4	7.0	-17.3
Central Asia	-5.7	-9.1	2.3	-6.2

* weighted by 1995 GDP   ** weighted by 1995 GDP, Albania 1997 growth rate assumed to be equal to 1995

Source: WDR, EBRD

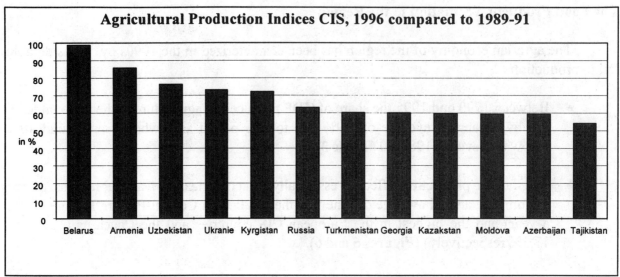

Figure 5

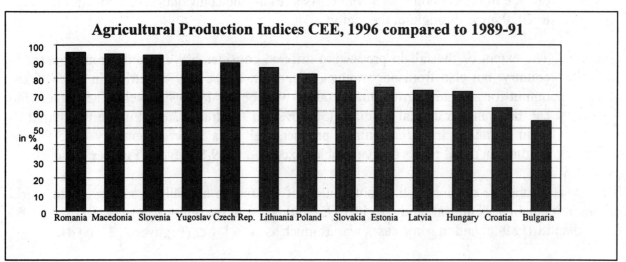

Figure 6

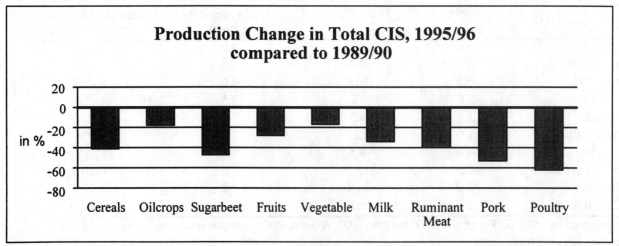

Figure 7

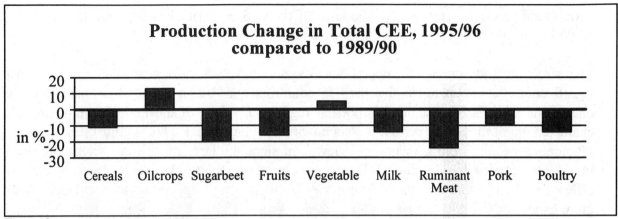

Figure 8

Yet behind the general decline, the first signs of recovery can already be discovered. In practically all the countries of Central-Eastern Europe the relatively rapid decline in production had already stopped in 1993-94 and from 1995 on, the sector's output was once again growing in almost all of the countries, even if at a moderate rate. This is the reason for the better performance in these countries than the five-year average for the CIS countries (only -1.9% a year). An upswing in agricultural production can also be observed in those few CIS countries (mainly the Caucasus region) where radical land reform was carried out. It is also worth mentioning that the level of agricultural production has failed to recover in the few countries (Belarus, Turkmenistan, and Tajikistan) where the basic elements of the planned economy are still in place. A further decline in output is not likely in the key CIS countries - Russia and Ukraine; however, over the short term a substantial upswing is not probable either.

As a result of this decline in agricultural output, the region's contribution to world production in all main products naturally fell and agricultural trade in the region was also transformed (**Figures 9, 10**, and **11**).

- Despite the decline in production, the region's share of world trade did not shrink substantially, and, in the case of some products, there was even an unquestionable increase. This was made possible (or required) by the fall in domestic consumption and by the new situation created with the disintegration of the Soviet Union.

- Some countries there has been considerable change in the composition of agricultural trading partners. The region's agrarian trade determined by the CMEA and basically built on internal relations has now given way to a wide opening towards other parts of the world. Conversely, growth in most CIS countries is seriously constrained by the collapse of traditional markets and failure to develop alternatives.

- However, while the structure of the region's agricultural exports and imports has changed considerably as a result of declining consumption, the balance of agricultural trade for the region as a whole did not deteriorate. Overall the region continues to be a net importer of agricultural products. The negative balance of the CIS countries has been reduced slightly, while in the case of Central-Eastern Europe it has grown.

- The structure and source of imports and exports have also changed. Perhaps the most significant structural change is that the CIS countries, and Russia in particular, have become one of the world's biggest meat importing regions. In place of the massive grain imports characteristic of the Soviet period, Russia now mainly buys meat. This is quite clearly a more favourable solution from an economic viewpoint since the large quantity of grain purchased in earlier decades by the Soviet Union was used in animal husbandry with very low efficiency. At the same time the CIS countries are increasingly appearing on world markets as grain exporters.

- The role of Central-Eastern Europe in the world meat trade had diminished, while in the early 1990s, the region's significance on grain markets has increased.

- Another important change is the growth in the importance of quality processed products in the region's agrarian trade, and together with this, in the share of the developed countries, especially the USA and the European Community in the region's food imports.

- Sales within the region continue to be of great importance for practically all countries. However, the competition from outside is making it increasingly difficult to sell products at Council of Mutual Economic Assistance (CMEA) level standards within the region. Some of the countries of Central-Eastern Europe are having increasing success on the markets of the developed countries, but for the majority of countries in the region, selling within the region remains practically the only possible direction for the export of their agricultural produce, often in the form of special barter deals.

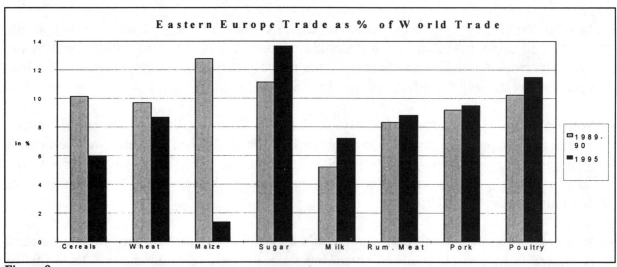

Figure 9

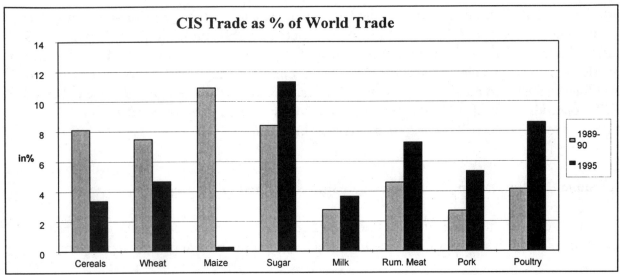

Figure 10

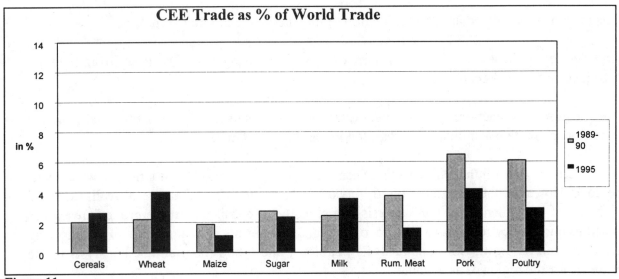

Figure 11

On the whole, the region's agrarian trade is becoming steadily integrated into the agrarian trade of the world and the European region. This process is most advanced in the case of the countries of Central-Eastern Europe where the CEFTA offers further possibilities for regional co-operation. In the great majority of the countries concerned, a liberal agrarian trade policy is also assisting in the integration of the countries of the region into world agrarian markets. Most of the countries in the Central-East European region are members of the WTO, or their admission is pending. The obligations accompanying the anticipated EU membership for most of these countries are also having a growing influence on their trade policy.

The agrarian trade policy of the CIS countries is changing and is a source of considerable uncertainty in the medium-term. The voices of those demanding increased protection of internal agrarian markets are becoming louder in the CIS countries, and especially in Russia. A proposal has also been elaborated for the introduction of a considerably projectionist "Common CIS Agrarian Trade and Support Policy," similar to that of the European Community. Ukraine wishes to restrict imports of animal products to 10% of its domestic output. It is not yet possible to clearly foresee the outcome of these processes, but it seems likely that the agrarian trade policy of the CIS countries will shift towards growing protectionism unless the protracted entry of Russia and Ukraine to the WTO sets limits to this unfavourable change.

What Stage Has the Reform Process in Agriculture Reached in the Region?

In 1990-91 the region set out on the path of creating market economies based on private property. In all countries the most important basic elements of the reform process have been:

- the liberalization of prices and markets, the creation of a market-compatible system of conditions in the macro agrarian economy;

- the privatization of land and transformation of the inherited economic structure;

- the de-monopolization and privatization of food processing and the trade in agricultural produce and capital goods; and

- the creation of a functioning rural bank system and establishment of the institutional structure and system of state administration required by market economies.

There has been little difference between one country and another in terms of what needs to be done. However, there are quite big differences when it comes to the pace of realisation and the manner of implementation. Our analysis intentionally emphasized "on the ground" results as opposed to pure policy reforms (which are often legislated but not implemented). The following conclusions can be drawn from this analysis (**Tables 4** and **5** and more detailed analysis in the **Section B**) for the region of as a whole[3]:

- The reform process is considerably more complicated and complex that anyone had originally expected.

- The results of the reforms have not yet met original expectations. The relatively rapid growth of production that characterized the Chinese reforms has not occurred. Transformation of the economic structure proved to be a far more complex task. This is due, largely, to the incomplete application of the basic element of farming, the private family farm. To a large extent the inherited large-unit structure has survived the changes.

---

[3] The description of the status of reforms for each country matrix was compiled by the World Bank staff most familiar with that country's agricultural policies. Numerical ratings were then assigned to each of the five reform categories in accordance with the criteria listed in **Table 5**. These ratings were then revised in several review sessions to improve consistency of rankings. An earlier version of this analysis was presented in Csaki and Lerman, 1997.

Table 4: Overview of the Status of Agricultural Reforms in CEE and CIS Countries (mid 1997)
1 = Centrally Planned Economy   10 = Completed Market Reforms*

Country	Price & Market Liberalization	Land Reform	Agroprocessing & Input Supply	Rural Finance	Institutional Framework	Total Score
Hungary	9	9	9	8	8	8.6
Slovenia	8	9	8	8	9	8.4
Czech Republic	9	8	8	8	8	8.2
Estonia	10	6	7	7	9	7.8
Latvia	7	9	7	7	8	7.6
Poland	9	8	7	6	8	7.6
Slovak Republic	7	7	8	8	7	7.4
Armenia	7	8	7	7	8	7.4
Lithuania	7	8	7	6	7	7.0
Macedonia, FYR	7	7	8	4	6	6.4
Albania	8	8	8	3	5	6.4
Georgia	7	7	5	6	6	6.2
Romania	7	7	6	6	4	6.0
Russia	7	5	7	6	5	6.0
Kyrgyz Republic	6	6	6	6	5	5.8
Moldova	7	6	7	5	4	5.8
Croatia	6	5	6	6	6	5.8
Kazakhstan	7	5	7	5	5	5.8
Bulgaria	6	7	5	4	5	5.4
Ukraine	7	5	7	5	3	5.4
Azerbaijan	6	6	5	4	4	5.0
Tajikistan	4	2	5	3	5	3.8
Uzbekistan	4	1	1	1	4	2.2
Turkmenistan	2	2	1	1	3	1.8
Belarus	3	1	2	2	1	1.8
Average Score	6.7	6.1	6.2	5.3	5.7	6.0

* An explanation of the numerical ratings is given in **Table 5**.
Source: World Bank Estimates.

Table 5: Key to Numerical Ratings Used in Table 4

Price and Market Liberalization	Land Reform	Privatization of Agroprocessing and Input Supply	Rural Financial Systems	Institutional Framework
1-2. Direct state control of prices and markets.	1-2. System dominated by large-scale farms.	1-2. Monopolistic state owned industries.	1-2. Soviet type system, with "Agrobank" as the sole financing channel.	1-2. Institutions of command economy.
3-4. Deregulation with indicative prices, and price controls; significant NTB on imports or exports.	3-4. Legal framework for land privatization and farm restructuring in place, implementation launched only recently	3-4. Spontaneous privatization and mass privatization in design of early implementation stage.	3-4. New banking regulations are introduced; little or no commercial banking.	3-4. Modest restructuring of government and public institutions.
5-6. Mainly liberalized markets constrained by the absence of competition and some remaining controls on trade policy.	5-6. Advanced stage of land privatization, but large-scale farm restructuring is not fully complete.	5-6. Implementation of privatization programs in progress.	5-6. Restructuring of existing banking system, emergence of commercial banks.	5-6. Partly restructured governmental and local institutions.
7-8. Liberal markets and fairly liberal trade policies with not fully developed domestic markets.	7-8. Most land privatized, but titling is not finished and land market not fully functioning.	7-8. Majority of industries privatized.	7-8. Emergence of financial institutions serving agriculture.	7-8. Government structure has been refocused while research, extensions, and education is being reorganized.
9-10. Competitive markets with minimal government intervention.	9-10. Farming structure based on private ownership and active land markets.	9-10. Privatized agro-industries and input supply.	9-10. Efficient financial system for agriculture, agro-industries, and services.	9-10. Efficient public institutions focused on the needs of private land market agriculture.

Source: World Bank Estimates

- The pace of transformation of the agrarian sector and the rural economy is lagging behind the rate of changes in the economy as a whole.

- Surprisingly, the biggest transformation has taken place in the price and market environment, while there is a substantial lag in solving the financing problems of agriculture and in the area of institutional reforms.

- There are very considerable differences in the reform performances of the individual countries. Due to the adoption of more comprehensive transition policies, the CEE countries are quite clearly far more advanced than the CIS countries.

- The transformation of agriculture is most advanced in Central Europe and, in particular, in the EU candidate countries. Even here, reform of the agrarian sector has not yet been fully accomplished. The results of our analyses agree with the EU evaluation in finding that further reforms are needed, principally in the area of the institutional system and in the financing of agriculture, but land reform and also the transformation of the inherited economic structure is still unfinished in practically all of the countries.

- The transformation of agriculture in the CIS countries is still in its early stages. Distortions continue in the production, pricing, and marketing of "strategic" products, and the system of institutions and instruments of the planned economy has not yet been fully dismantled.

Many valuable conclusions can be drawn from the analysis of the experiences of the countries leading in the transformation. The following can be stated:

- The general economic upswing can greatly assist the agricultural reforms. The greatest progress has been made in transformation of the sector by those countries where the general economic recovery has also begun.

- Development in the non-agricultural segment of the rural economy is of key importance for the recovery of agriculture. In the great majority of the countries that are the most advanced in the reforms, it has been the upswing of the rural economy surrounding agriculture that has made possible a substantial reduction in the numbers of people employed in agriculture, and at the same time, an improvement in the efficiency and competitiveness of agriculture itself.

- An important factor in the successes achieved in reforms is consistency, and the combined implementation of parallel steps in areas related to the reforms.

- The greatest reform progress has been made by those countries that are reforming in very large steps, despite the great difficulties that these efforts are causing greater difficulties in the short-term. In most cases the appeals for a gradual approach appear to be a sign of the lack of will; this is especially the case in the CIS countries.

In all countries the process of agricultural reforms has been strongly influenced by day-to-day politics. Very often, politics have been and still are determining the pace and extent of reforms, at the expense of economic rationality. In general, there is a lack of a carefully considered, long-term strategies, and an objective and realistic evaluation of the economic consequences of the different possible solutions. As a result, the negative economic consequences of the transformation are greater than the unavoidable minimum, even in the most advanced countries.

The situation in the different main areas of transformation covered in our analysis can be summarized as follows:

a) *Liberalization of prices and markets.* Practically all countries have taken major steps in this direction.

- In Central-Eastern Europe, the macro-economic environment for agriculture that is characteristic of market economies has been developed. The prices and the system of regulations are open, more or less, to world market influences. Support for agriculture and protection of internal markets has become stronger than in the first half of the 1990s. However, the level of indirect and direct supports for agriculture remains below the average for the European Union. Frequent changes made in the system of regulations represents a problem. It remains to be seen what strategy the countries preparing for EU membership will opt for to introduce the EU's Common Agricultural Policy. It is a positive trend that methods directly serving efficiency and competitiveness are increasingly coming to the fore in the system of agricultural supports.

- State intervention remains strong in the CIS countries, in both price formation and trade policy. In these countries, despite the frequently proclaimed direct support for agriculture, the agrarian sector suffers serious losses due to the price policy and trade restrictions (especially export controls and taxes) separating it from world markets. Calculations show that the balance of the different interventions is negative for agriculture. It would appear that governments are trying to make agriculture continue to bear the burden of providing cheap food for the urban population. The growing intervention of regional authorities in the functioning of the agricultural sector is a relatively new phenomenon. This can be observed particularly in Russia and Ukraine.

b) *Privatization of land, reorganization of the large farm units.* Land reform and land ownership continues to be the subject of heated debates in practically all countries of the region.

- In the countries of Central-Eastern Europe the privatization of land based on some form of compensation is largely approaching completion. The new farm structure is characterised by a varied mix of small and large units. The remaining pieces of the state owned units from the socialist period are also increasingly undergoing change and adapting to market economy conditions. The legal settlement of land ownership relations is not yet completed, and the land register and the emergence of a market for land are still in the initial stages. In a few countries a heated debate is being conducted on the ownership of land by companies and foreign nationals.

- While land has formally passed into private ownership in the decisive part of the CIS countries (Russia, Ukraine, Belarus), the large-unit sector remains practically untouched. The co-operatives and state farms have been transformed into share companies (Brooks and Lerman, 1994 and Csaki and Lerman, 1997). However, in practice there has been no change in the real questions of substance. These large units continue be controlled by a central management structure and operate with low efficiency and face increasing financial problems. The role of independent private farming is marginal, not least of all because of the deterrent effect of the undeveloped market relations. At the same time, radical land reforms have been carried out in a few countries of the former Soviet Union, as a consequence of the special political and economic situation. This is the case for Armenia and Georgia where independent private farming now dominates. Here, the distribution of land carried out on the basis of family size resulted in very small farm sizes and this has gone together with a steep decline in agricultural production for the market. Conversely, in some CIS countries (Uzbekistan and Tajikistan) private ownership of land is prohibited by the constitution and the lease hold arrangements are an added uncertainty.

c) Privatization of the food industry and of trade in agricultural produce and capital goods for agriculture. Formally, very substantial progress has been made. However, the food industry, which is technologically backward and incapable of quality production, is one of the biggest obstacles to the further development of agriculture in the region.

- In Central-Eastern Europe, with the exception of the Baltic states, privatisation of the agricultural environment has been carried out in keeping with the principles of the privatisation in general, and for the most part is nearing completion. A lag can be observed in Romania, Bulgaria and the countries of ex-Yugoslavia. A special case is Hungary, which, as a result of the liberal privatisation process, now has perhaps the most developed food industry of the region, due also in considerable part to substantial foreign investment.

- The CIS countries opted for what, on the whole, is a less effective solution for privatization of the food industry and agricultural capital goods and commodity purchasing. In the course of privatization, unlike the other areas of the economy, priority was given to agricultural producers, giving them majority ownership of these branches, on special terms or entirely free of charge. Contrary to expectations, this solution did not result in new, well capitalized owners and more favourable conditions for agricultural producers. In fact, the technological decline of the food industry accelerated and because of the complicated ownership structure it became extremely difficult to involve foreign capital. Further steps and the renewed settlement of ownership are required to set the food industry on its feet. Instruments for this could be new liquidation procedures and support for new investments. Unfortunately, the widespread bureaucracy and corruption in the CIS countries seriously obstruct both the further transformation and the establishment of new plants and small businesses.

d) Agrarian financing. This is one of the most critical areas for the agrarian sector in the region.

In the CEE countries, after intensive preparations and a lengthy transitional period, the financing of agriculture has improved considerably since 1994, although the new private institutions are managerially weak and financially vulnerable. This is the result, partly of the reforms implemented in the banking system, and partly of the credits extended by the gradually recovering food industry and the agrarian trade and capital goods supply. Finally, the creation of an agriculture-oriented rural banking system has also been progressing, resulting in the establishment and increasingly active operation of agricultural credit co-operatives and financial institutes specialising in rural areas.

- In the great majority of CIS countries there is practically no functioning rural financial system similar to that in developed countries. The inherited banking system continues to provide financing for large plants using the accustomed methods of the earlier period (state credit, financing the supply of produce for state stocks), but there are practically no provisions at all for the financing of the private sector. The beginnings of a system of agricultural credit co-operatives have appeared in the countries most advanced in the transformation of agriculture, namely Armenia, Georgia and recently also Moldova, and the credits extended by the processing industry are also growing.

e) Institutional reforms. Transformation of the institutional structure is proceeding more slowly than practically all other areas of reform throughout the region.

- Institutional reforms have accelerated in Central-East Europe since 1995, stimulated by the challenges of EU accession. Despite these tangible developments, the institutional system of agriculture requires substantial further transformation in these countries. In addition to the modernisation and reform of state administration, further qualitative development is required in practically all areas of the institutional systems for market agriculture, including consulting, training, and research (see also EU 1997).

- Apart from minor changes, the institutional system of the former centrally planned economy continues to operate in the CIS countries and continues to act as a serious curb on the transformation of the sector. Due to the general economic recession, there are fundamental disorders in the operation of the institutional system and the underpaid and unmotivated state bureaucrats strive to supplement their incomes through corruption. Training and research centers are suffering from severe financial problems and in some countries they receive no financial support from the government budget.

The Prospects for Development in the Region

Three factors will directly determine the long-term course of agricultural production in this region, namely: a) the investments and current assets available; b) the rate of technological renewal, together with the pace of introduction of new advances in biology and higher-yield crop and livestock strains; and c) changes in organisation and the human environment of production. However, the trends in these three factors will depend above all on the further fate of the reforms, the results achieved in the transformation of agriculture, and on the general development of the economy of the countries concerned and, as a function of this, on the internal demand for agricultural produce.

In many CEE countries the process of sector transformation will probably be completed within the next four or five years. This, in itself, will improve the conditions for directly determining agricultural production, and together with the anticipated acceleration of general economic development, could lead to the stabilisation of food production, more efficient production, and a greater upswing driven by potential comparative advantages. Although the effect of EU accession cannot be clearly quantified, it will probably give further stimulus for the growth of production in the countries concerned.

It is far more difficult to predict changes in the CIS countries. It seems probable that further difficult years lie ahead for the sector in this region, compounded by the struggle between conservative and progressive forces. The reforms will probably continue to advance only slowly. However, it is not likely that the decline in production will continue. The results of 1995-96 and the forecasts for 1997 indicate that the fall in the region's agricultural output has reached its lowest point. In the short-term the most likely course is moderate growth alternating with stagnation. Substantial growth in production can only be expected if substantive progress is made in the reforms that directly determine the factors of production. This will probably happen only in the medium-term.

Many studies have forecast development in the region as a subset of global projections.[4] These calculations, and our analysis of the progress of sectoral reforms, suggest that:

- in the CEE countries development can be expected to follow a production course close to, or higher, than the world average. For a few products this trend could be even more favorable.

- In the CIS countries the probable course of development will be more modest, and probably below the world average. Naturally, in this group of countries there will be exceptionally big differences between the individual countries. Trends at region level will obviously be determined by the results in Russia and Ukraine where, in our judgement, the conditions for a bigger upswing will probably not be created until the final stage of the forecast period in the best of cases.

- In general, crop production will begin to expand sooner than animal husbandry. In some countries, especially Ukraine and Kazakhstan, the ecological and technical potential is there for a quick recovery in the crop sector, mainly in grain production. Such a quick recovery, however, assumes the immediate implementation of appropriate sector reforms, which in our view do not appear very likely.

In Central-Eastern Europe animal husbandry will probably continue to stagnate while in the CIS countries it is highly likely that the decline will continue for most of the period under analysis.

---

[4] See Agriculture Canada (1996), European Commission (1997), FAO (1997), FAPRI (1997), IFPRI (1996), OECD (1997), USDA (1994).

- The increase in quality demands will probably outpace reconstruction of the food industry. In general, it can be said that the backward food industry could restrict the growth of produce output for a relatively long time to come.

- The difficulties of the transformation and the problems of the food industry will probably strengthen the protectionist trends in the CIS countries. However, it is to be hoped that consumer interests will prevent any serious restriction of imports.

In the case of *grains,* a performance around the level of full self-sufficiency can be expected, with a small volume of exports likely in better years and smaller net imports when weather conditions have been unfavorable. Only minor shortfalls are likely for *wheat*, while there will probably be a few million tons of surplus *maize*, mainly from the output of the Central European countries. The region will probably remain a moderate importer of *sugar*, with the CEE probably regularly producing a small surplus and the CIS countries a deficit of 2-3 million tons a year. There will be a constant surplus on the *milk* market in the Central European countries but substantial net imports can be expected for the CIS countries. In the case of *beef* and *mutton* both Central-Eastern Europe and the CIS will probably remain net importers with the volume of imports likely to stabilise around 1.5 million tons a year. The *pork* surplus of the CEE countries seems set to increase, while the CIS countries will probably still be in a moderate import position in 2010. In the field of *poultry*, the trends will be similar to that for pork: a growing surplus in the CEE countries and a long-term, although diminishing, demand on the markets of the CIS countries.

Considering the main products, the prospects for the region as a whole are that the net imports will fall slightly even if the forecast moderate general growth of agriculture is realised. However, the region as a whole will continue to be a net food importer. It would appear that exploitation of the very considerable natural resources will remain in the more distant future and that up to the year 2010 the region will not become a substantial source for satisfying the growing global food demand. Naturally, because of differences in the production levels of the individual countries and in the pace of reforms, differences within the region will grow.

References

Agriculture Canada. 1996. *Medium Term Baseline Projections 1996*. Ottawa.

Brooks, Karen, and Zvi Lerman. 1994. "Land Reform and Farm Restructuring in Russia." *World Bank Discussion Paper 327*. Washington, DC.

Csaki, Csaba. 1995. "Where is Agriculture Heading in Central and Eastern Europe?" *Agricultural Competitiveness*. Dartmouth, NH.

Csaki, Csaba. 1996 "Agriculture Related Issues of EU Enlargement to Central and Eastern Europe." *Society and Economy in Central and Eastern Europe*. 17(1):35-46.

Csaki, Csaba, and Zvi Lerman. 1997. "Land Reform in Ukraine: The First Five Years." *World Bank Discussion Paper 371*. Washington, DC.

Csaki, Csaba, and Zvi Lerman. 1997. "Land Reform and Farm Restructuring in East Central Europe and CIS in the 1990s: Expectations and Achievements after the First Five Years." *European Review of Agricultural Economics*. 431-455.

Csaki, Csaba, and Zvi Lerman.1996. "Agricultural Transition Revisited: Issues of Land Reform and Farm Restructuring in East Central Europe and the Former USSR." *Quarterly Journal of International Agriculture*. 35(3):211-240.

European Commission. 1997. *Agenda 2000*. Brussels.

European Commission. 1995. *Agricultural Situation and Prospects in the Central and Eastern European Countries*. Brussels.

FAPRI. 1997. *International Agricultural Outlook*. Iowa State University/University of Missouri. Columbia, MO.

FAO. 1997. *The State of Food and Agriculture, 1996*. Rome.

International Policy Council on Agriculture, Food, and Trade. 1997. "Agriculture and EU Enlargement to the East." *Position Paper 4*, Washington, DC.

Nurul Islam, ed. 1995. *Population and Food in the Early Twenty-First Century*. International Food Policy Research Institute. Washington, DC.

OECD. 1997. *The Agricultural Outlook, OECD*. Paris.

USDA. 1994. *The Former USSR: Situation and Outlook Series.* International Agriculture and Trade Reports, Washington, DC.

World Bank. 1997. *The State in a Changing World - World Development Report.* New York: Oxford University Press.

B. Country Analysis

Central and Eastern Europe

Policy Matrices

ALBANIA

Total Population	3.2 million	Agriculture in GDP (1995)	55.8%	Agricultural output in 1995	
Rural Population	59.4%	Food and agriculture in		As percentage of 1990 level	124.4%
		Active labor force (1995)	65.0%	Livestock production 1995	
Total Area	2.8 million ha	Food and agriculture		As percentage of 1990 level	153.7%
Agricultural Area	1.13 million ha	in exports (1995)	86.1%	Share of livestock in agriculture	
		Food and agriculture		(1995)	47.1%
Arable Land	62%	in imports (1995)	28.4%	Agricultural area in private	
Irrigated	7.1%			Use (1995)	100%
Orchards	17.6%	Currently a net exporter of:		Share of family farms in total agricultural	
Natural Grassland	37%	medicinal plants, fish and fish		land (1995) 100%	
Forested	37.5%	products and tobacco		Share of private sector in total	
				Agricultural output (1995)	100%

ISSUE	STATUS OF REFORMS	OBJECTIVES PROPOSED ACTIONS
1. Macro-Economic Framework for Agriculture A. Prices/Subsidies	**Markets and prices are fully liberalized with minimal price distortion, but market structures are weak and poorly integrated** • All agricultural producer and consumer prices deregulated (bread price controls removed mid 1996). • Government has retained a floor price (below parity) for wheat and small subsidies for irrigation, and flour mills and bakeries • General State Directorate for State Reserves still active in wheat market and a potential source of price distortion • Markets remain weak due to poor infrastructure and a very low marketed surplus • Well developed public transfer system for low income groups in rural areas but has limited resources	**Distortion free marketing and incentive system and a more open trade policy** • Remove floor price for wheat and remaining subsidies • Limit activities of GDSR to the acquisition, management and disposal of a small strategic reserve, which does not distort domestic prices • Support development of infrastructure and market centres • More effective means to target transfer payments to rural poor
B. Trade Policies	• WTO Accession in progress • Import restrictions: ad valorem tariffs of 10-40%, underpinned by reference prices • Increasing pressure to raise tariffs. • Private sector took over cereal imports from GDSR in late 1996 • Numerous bilateral trade agreements but none afford preferential access for agricultural exports	• Complete accession to WTO • Resist pressure to raise tariffs and continue efforts to reduce and rationalize overall levels of protection • Additional multilateral and bilateral trade agreements
C. Taxation	• VAT introduced mid 1996; 22%, no exemptions. • Land tax suspended prior to elections in 1996	• Full implementation of VAT

ISSUE	STATUS OF REFORMS	OBJECTIVES PROPOSED ACTIONS
2. Land Reform and Farm Restructuring	**Comprehensive re-distribution of all public agricultural land completed, but most of this land still lacks secure, unambiguous property rights** • All agricultural land re-distributed in 1991-92 • Ownership certificates (tapi) issued for 60% of ex-cooperative land but full registration and titling for only 3% (15,000 ha) • Owners of ex-state farm land only have use rights, not ownership rights • Forest and pasture land still under public ownership • Inadequate legal and administrative procedures for resolving ownership disputes • Land sales prohibited until July 1995, but land market still not operating (1997) due to the small percentage of land with titles, and inadequate administrative procedures for registration and transfer	**Full ownership rights and a working land market for fair and efficient land transfer** • Complete issue of tapis and land titles • Confer full ownership rights for ex-state farm land • Transfer forest and pasture land to village control • Resolve ownership disputes • Develop administrative procedures for registration and titling of land
3. Competitive Agro-Processing and Services for Agriculture	**Privatization in progress** • Two-thirds of agro-processors now privatised, recent government commitment to accelerate and complete privatization of the remaining (mostly larger) State-Owned Enterprises • Limited foreign investment in privatization process • Newly privatized agro-processors hampered by outdated equipment, lack of credit and a weak understanding of competitive markets • Product quality is low and new systems for setting and monitoring quality standards are not yet in place • Seed sector still dominated by public agencies and now virtually inoperative, with severe consequent seed shortages • Small but dynamic system of private sector input dealers has evolved but they are constrained by poor access to credit	**Competitive, privately owned agroprocessing, input supply and service subsectors operating with minimal government protection** • Complete privatization of remaining State-Owned Enterprises • Increase access to credit for private sector enterprises, and continue support for improving management techniques • Complete establishment of new systems for defining and monitoring product quality • Reform legislation and institutional structures for seed certification and testing; and promote private sector import and multiplication of seeds

ISSUE	STATUS OF REFORMS	OBJECTIVES PROPOSED ACTIONS
4. Rural Financing	**Lack of an appropriate financial system for agriculture**	**Viable financial institutions serving the agricultural sector efficiently**
	• Rural credit is provided only by a State-Owned bank (Rural Commercial Bank) which is technically insolvent due to low loan recovery	• Privatize and re-structure Rural Commercial Bank
	• There are very few commercial banks and none are interested in agriculture	• Develop legal framework for loan enforcement
	• Continued shortage of credit and high interest rates	• Expand ADF where appropriate
	• Slow progress with privatization, rehabilitation and re-structuring of State-Owned Banks	• Develop legal framework for small-scale rural banking and set up Pilot projects
	• Donor funded Agricultural Development Fund (ADF) provides credit for rural infrastructure and small enterprises in poorer mountain areas	
	• No credit cooperatives or savings and credit associations	
5. Institutional Framework	**Public institutions not yet adjusted to the needs of small-scale private farmers; lack resources and trained personnel**	**Efficient and effective public sector administration and support for commercial private agriculture**
	• MAF now has mainly regulatory functions, but has yet to re-organize and train personnel to implement these functions effectively	• Continued support for re-organization and staff training
	• Agricultural extension services are weak and reform of the agricultural research system has yet to begin	• Continued support for reform of extension and research programss
	• Limited MAF capacity for policy analysis	• Continued efforts to improve the capacity for policy analysis
	• Veterinary services privatized except for regulation and border control	

BOSNIA AND HERZEGOVINA

Total Population (1996)	3.6 mil.	Food and agriculture in GSP 1990	14 %	Agricultural output in 1991 as percentage of 1990 level	93 %
Rural Population	50 %	Food and agriculture in active labor (1990)	18 %	Agricultural output in 1995 as percentage of 1990 level (estimate)	33 %
Total Area	5.2 mil	Food and agriculture in exports (1996)	n.a.	Livestock production in 1995 as percentage of 1990 level (estimate)	8 %
Agriculture area:	2.5 mil ha.	.. in imports (1996)	n.a.	Share of livestock in agriculture (1991)	44 %
Arable land	63 %	Traditionally net exporter of: livestock products		Share of livestock in agriculture (1995)	26 %
Orchards	4 %	(dairy products, meat),		Share of independent private farms in total arable area (1991)	94 %
Irrigated	0.3 %	fruits and vegetables, and		Share of private sector in total agricultural output	
Forested	46%	wine		(1991)	75 %

ISSUE	STATUS OF REFORMS	OBJECTIVES PROPOSED ACTIONS
1. Macro-economic Framework for Agriculture	**Market liberalization is largely a consequence of the war; however, introduction of Government intervention is being envisaged.**	**Competitive and functioning agricultural markets without Government intervention.**
A. Prices/Subsidies	• Agricultural producer and consumer prices are free, generally due to the lack of financial resources and administrative capacity of the Government following the recent war.	• Maintain fully liberalized prices. Unify markets between the FBH and Republika Srpska, establish freedom of movement of goods.
	• No known Government procurement quotas.	• Guarantee that farmers are able to retain profit for development.
	• Domestic grain price adjusted to border price, no known controls of bread prices.	• Avoid introduction of government intervention.
	• Pressure to introduce Government producer price supports in the Federation of Bosnia and Herzegovina (FBH) to facilitate recovery of war-affected food production industries.	• Consider an adequate "social safety net" of subsidies targeted to low income and vulnerable consumers.
	• Domestic producer prices in the FBH adapted to border prices. Producer prices in Republika Srpska estimated below border prices due to lack of marketing competition.	• Introduce price information systems for transparent and timely access to cost changes by all concerned.
B. Trade Policies	• Average applied import tariff in the FBH is 13%, in Republika Srpska 28.5%; slightly lower tariffs for agricultural products.	• Remove remaining administrative and quantitative restrictions on exports and imports. Reduce import tariffs to a low and uniform rate, apply the same tariff rate in both Entities of Bosnia and Herzegovina.
	• Special trade arrangements with Croatia (for the FBH) and with the Federal Republic of Yugoslavia (for Republika Srpska) are in place where only a 1% import tax is applied.	• Proceed with demonopolization, corporatization, and phased privatization of internal and external trading enterprises.
	• New State-levels laws are planned that would liberalize and harmonize trade policies of the FBH and Republika Srpska.	• Pursue active trade policy to improve market access for Bosnian food and agricultural products, especially in former Yugoslav republics and in Central and Eastern Europe.
	• Meanwhile, FBH is considering to introduce export subsidies for milk, tobacco, and wheat.	

ISSUE	STATUS OF REFORMS	OBJECTIVES PROPOSED ACTIONS
C. Taxation	• Tax burden on farmers very low, farmers operate in informal sector. • Land taxes existed prior to the war, are currently not collected, future Government intentions are unclear. • Taxation of food processing is in line with general taxation of businesses. Very high social charges on dependent labor.	• Larger private farms should be incorporated into the regular business tax system.
2. Land Reform and Farm Restructuring	95% of farm land in private ownership, but land consolidation and clarification of land ownership rights urgently needed. • Agriculture was de facto decollectivized during the 1991-1995 war. • Federation Government plans to re-establish farmer associations based on voluntary principles (draft Cooperative Law has been prepared). • Only about 5% of arable land is held by State farms. • Private farmland holdings average 3-5 ha, often fragmented into several plots. Former Yugoslav laws limiting land ownership to 10 ha in flat lands and 15 ha in hilly areas have been abandoned. • Pastures and meadows remain state and municipality owned. • Land sales are allowed; however, market for land and leasing is developing slowly. • Despite the war, the technical conditions (titles, registration system) for a functioning land market are generally in place. However, ownership rights are heavily disputed due to the large share of displaced persons and the growing population of returning refugees.	**Individual private farms as the dominant component of the farming system with secure and transferable ownership rights.** • Restore and further develop the war-damaged land registration, information and cadastre system to provide security of tenure, full information on land transactions, and a basis for land taxation. • Implement a program to promote the emergence of land markets to support land consolidation and the move towards a more efficient holding structure. • Establish competitive land mortgage and credit systems. • Provide technical assistance to farmers to achieve optimum production mix and efficiency. • Speed up the work of the Commission for Real Property Claims, set up under Annex 7 of the Dayton-Paris Peace Agreement, so as to address disputed land ownership rights. Issue temporary land user rights in the interim.

ISSUE	STATUS OF REFORMS	OBJECTIVES PROPOSED ACTIONS
3. **Competitive Agroprocessing and Services for Agriculture.**	**Delays in privatizing agroprocessing and services for agriculture, but generally strong commitment for rapid privatization.** • Disputes prevail over responsibility for enterprise privatization: State vs. Entity responsibility. Draft privatization laws at the Entity levels are ready. Resolution of the dispute expected by mid-1997. • Privatization of agroprocessing and input supply would be part of the overall privatization program. There are only about 10 State farms left in the country to be privatized. • Foreign participation in the agroprocessing privatization will depend on prospects for peace consolidation. Foreign participation actively encouraged by all governments. • Political risk insurance scheme in place for foreign input suppliers and investors.	**Competitive privately owned agroprocessing and input supply.** • Fully implement program of corporatization and privatization of the agricultural input supply, output marketing, and agro-processing enterprises. • Establish feasible and reasonable quality and safety standards for agricultural imports and exports. • Acquire technical assistance and training in enterprise management. • Promote joint ventures to tap foreign expertise, technology, capital, and provide access to foreign markets. • Promote research and development of new products and markets.
4. **Rural Financing**	**Lack of an appropriate financial system for privatized agriculture.** • Financing in agriculture is not adjusted to the needs of a market based privatized agriculture: State-owned banks are largely insolvent; new private banks lack the capital, deposits and expertise do deal with small-scale rural lending. • Restructuring of UPI Bank (agricultural lending) and the banking system in general not yet started. • No Government subsidies for rural/ agricultural lending. • Establishment of an NGO-based micro-lending program, supported by donors, is under way with good prospects to assist farmers. The objective is to establish rural credit cooperatives. • Donor-financed credit lines for small and medium businesses, including agro-processing firms, are in place and are used actively. Also in place are donor-financed small business advisory services, but limited to urban areas.	**Viable financial institutions serving the agricultural sector efficiently.** • Support restructuring and privatization of State banks; support new private banks. Reform banking supervisory framework. • Do not introduce fiscal means involving financial institutions to sustain the operation of state enterprises critical to food security. • Establish legal framework and implement program to organize local savings and credit societies. • Establish programs to assist farmers in farm financial management. • Promote the emergence of competitive insurance services for agriculture.

ISSUE	STATUS OF REFORMS	OBJECTIVES PROPOSED ACTIONS
5. Institutional Framework	**Adjustment to the new Dayton structure of government is slow and constrained by budgetary difficulties.** • Two new Ministries of Agriculture were created in the FBH and in Republika Srpska, reflecting the sector responsibilities assigned under the Dayton/Paris Peace agreement to each of the Entities. • Establishment of agricultural extension services planned in the Federation for mid-1997, donor support sought. No known plans for Republika Srpska. • Agricultural research and education suffered heavily under the war and is in need of support. Government-financed research and education in agriculture are seriously hampered by budgetary constraints.	**Efficient and effective public sector administration and support for commercial and private agriculture.** • Complete the establishment of the two new public agricultural administrations in the Entities, as well as at the regional level in the Federation (Cantons). Limit administration size as much as possible. • Fully de-link the Government from agricultural production, planning and management. • Initiate reconstruction and reform of agricultural research and education to make it needs-based and demand-driven. Introduce cost recovery by charging service fees to farmers.

BULGARIA

Total Population	8.5 mil.	Food and agriculture in GDP 1995		Agricultural output in 1995 as percentage of 1990 level	70%
Rural Population	31%		18%	Livestock production in 1995 as percentage of 1990 level	52%
Total Area	11.1 mil ha.	Food and agriculture in active labor (1995)	18%	Share of livestock in agriculture (1995)	38%
Agriculture area:	6.2 mil ha.	Food and agriculture in exports (1995)		Agricultural area in private use (1995)	n.a.
Arable land	44%	in imports (1995)	20%	Share of independent private farms in total arable area (1995)	12.0%
Orchards	3.2%		10%		
Irrigated	28%	Traditionally net exporter: tobacco, sunflower seed, dairy products, wine, fruits and vegetables.		Share of private sector in total agricultural output (1994)	n.a.
Forested	30%				

ISSUE	STATUS OF REFORMS	OBJECTIVES PROPOSED ACTIONS
1. Macro-economic Framework for Agriculture	Economic crisis caused by hyperinflation and failure to institute structural reform came to a head in early 1997, when the old government resigned. While most prices and trade were liberalized in 1992, the government has sought to restrain some prices (e.g., grains, tobacco, wood) by trade controls and export restrictions in order to protect domestic industries and urban consumers.	Distortion free market and incentive system.
A. Prices/Subsidies	• Previous system of price and profit margin control was ended by new government, but replaced by a less restrictive, but still counterproductive system of "negotiated prices," with negotiations overseen by government • Farm gate prices were the lowest in Europe and at least 50% below world market levels partly due largely to trade restrictions and partly due to widespread racketeering, which was also facilitated by the trade controls. Incoming government has tried to reverse the anti-agricultural discrimination, and has set a support price much too high for wheat.	• Eliminate price and trade restrictions and impediments, especially support taxes and licensing. • Ensure law and order and fair competition in the countryside. • Replace the central-planning type "Law for Protection of Agricultural Producers" by market-oriented legislation. • Eliminate the negotiated price system.

ISSUE	STATUS OF REFORMS	OBJECTIVES PROPOSED ACTIONS
B. Trade Policies	• Incentives to produce have been undermined by capricious manipulations of trade regulations by the government. Temporary duty-free import quotas are used in ad hoc manner to reduce domestic prices, which undermines incentive of private sector to fill shortages. • Export ban on grain and grain products and oilseeds and products was in effect through 1996. Estimated cost of this measure in exports foregone was at least US120 million annually. The new government replaced this with an export tax, which was, however, set at a prohibitive level. This was reduced, and eliminated at the end of 1997. • Export taxes on other farm commodities are in effect. • Some export items require a non-automatic license (e.g., live animals, seeds). • Imports and exports of many sensitive products are under an "automatic licensing" system in order to monitor domestic supplies. If domestic supplies or prices are judged to be inappropriate, trade controls can be re-imposed, or duty-free import quotas granted. • Import tariff regime provides fairly high and non-uniform protection, especially for processing industries. Some tariffs, including fertilizer, are high. • Emphasis on joining CEFTA	• Replace frequent ad hoc government interference in the foreign trade for agricultural products with a stable and transparent regime consistent with WTO principles. • Aggressively liberalize policies to undermine and eventually to eliminate special interest and corruption in foreign trade. • Eliminate taxes on all agricultural exports; • Eliminate discretionary import duty exemptions • Eliminate non-automatic and automatic licensing for all agricultural exports; • Reduce tariffs on fertilizer imports • Adopt a lower and more uniform tariff structure • Do not let plans to join CEFTA interfere with lowering of tariffs toward other trade partners as well.
C. Taxation	• Single 22% VAT also applies to food in general, but there are numerous exceptions.	• End VAT exemptions, particularly on flour and dairy products

ISSUES	STATUS OF REFORMS	OBJECTIVES PROPOSED ACTIONS
2. Land Reform and Farm Restructuring	**The transition from collective to private farming has stalled and remains unfinished.** • Initiated in 1991, the protracted process of land restitution and liquidation of collective farms has devastated most farm assets. • Frequent changes in the regulatory framework for land restitution have confused both administrators of the process and those seeking restitution. • About 50% of farmland has been restituted with land titles, however, data are uncertain as much higher and lower figures have often been quoted. • Since 1991, "collectives in liquidation" are a widespread business form in primary agriculture. • The land market is dormant, although short-term land leases are widespread. • The new government is committed to finishing the restitution of farmland and extending it to forest lands, but the process remains stalled due to lack of funding and some institutional and regulatory obstacles.	**Accelerate land restitution and liquidation of collective farms.** • Promulgate a national cadastre law based on best practices successfully tested elsewhere. • Promulgate a land lease law based on best practices successfully tested elsewhere. • Complete land registration, information and cadastre to guarantee security of tenure and information on land transactions. • Promote an active land market and the use of land as collateral to improve access to long-term credit. • Eliminate limits on maximum holding size and permit land purchases by foreigners. • Maintain neutral policy toward all legal forms of business association. • Complete liquidation of collective farms by auctions of remaining non-land assets. • Promote improved utilization of irrigation systems by transferring ownership to users.

ISSUE	STATUS OF REFORMS	OBJECTIVES PROPOSED ACTIONS
3. Competitive Agroprocessing and Services for Agriculture.	**Policy drift in privatization of agroprocessing and services has brought the sector into a deep crisis.** • Until recently implementing agencies avoided privatization decisions, although an adequate regulatory framework has been in place since 1992. • Strong intricate links exist between managers of state-owned enterprises and private firms. Transfer prices shift profit from state to private firms who control 80% of domestic and foreign trade revenue. • State-owned agro-industries are deeply in debt estimated at about 8% of GDP (1995). • By late 1997, about 26% of long-term assets of agriculture and food industry enterprises were privatized. The achievement of 80% by the end of 1998, the intended goal, is doubtful • A number of start-up companies in the food processing industry went bankrupt and some foreign investors have threatened to back out, mainly due to unstable and insecure business environment.	**Competitive, private agroprocessing and input supply in a stable macroeconomic environment.** • Expedite privatization and or liquidation of state-owned enterprises. • Include many SOEs currently not in the privatization process, including the state grain company, forest and lumber milling enterprises, and cadastral survey firms. • Establish a stable enabling environment. • Pass legislation supporting formation of participatory service cooperatives • Improve market information systems and external trade infrastructure • Develop warehouse receipts system • Change privatization practices to expedite program

33

ISSUE	STATUS OF REFORMS	OBJECTIVES PROPOSED ACTIONS
4. Rural Finance	**The banking system virtually collapsed as hyperinflation and dollarization of the economy spread.** • High risk stemming from uncertain property rights, low profitability and poor credit history greatly limit access to credit. • Avoiding banks, virtually all working capital needs are met from farmers' own-sources. • Limited subsidized credit has been allocated to well-connected parties. • Donor supported credit co-op network is small, but successful.	**Restoration of profitable conditions to farming as the first essential condition for improved access to rural credit; integration of the rural financial system within the overall financial system** • Restore confidence in the Bulgarian economy by a macroeconomic stabilization and banking reform. • Restore profitability to agriculture by eliminating price and trade restrictions, and avoid ad hoc interference in commodity pricing. • Restore confidence in Bulgarian banks by financial sector reform and effective bank supervision. • Eliminate interest rate subsidies because they distort financial markets and, if granted, would need to be rationed.
5. Institutional Framework	**Agricultural institutions including research and education have drifted into irrelevance, destroyed by policy drift, brain drain and inflation.** • Instead of guiding agriculture to a market-based system, the Ministry of Agriculture and Food Industry has focused only on ad hoc actions often motivated by a crisis or pressure from a special interest lobby. • Even basic statistical information is unreliable. • Research establishment has been decimated. Financial resources barely cover salaries and are spread too thinly over rapidly declining number of centers, staff and projects. Little innovative research is underway, both technology and equipment are outdated. • Adjustment in agricultural education has been limited and Public extension service does not exist.	**Start with a narrow focus on essential priorities in public sector administration to ensure success.** • Develop a solid information database for agricultural policy decision-making. • Rationalize agricultural services by salvaging remaining valuable assets in research and education. • Cut losses and dispose of assets with no prospects for becoming profitable again such as most state-owned livestock farms. • Transfer assets to users, such as irrigation infrastructure, buildings, and equipment.

CROATIA

Total Population	4.67 m	Agriculture in GDP (1995)	11.4%	Agricultural Output in 1994		
Rural Population	37%	Food and Agriculture in		as percentage of 1990 level		82%
		Active Labour Force (1995)	8.7%	Livestock Production 1994 as		
Total Area	5.65 mil ha	Food and Agriculture		Percentage of 1990 level		69%
Agricultural Area	3.21 mil ha	in Exports (1995)	11.3%	Share of livestock in agriculture (est)		45%
Arable Land	63%	Food and Agriculture		Agricultural Area in Private Use (1994)		67%
Orchards	4.4%	in Imports (1995)	12.4%	Share of Independent Family Farms in		
Natural Grassland	36%	Traditionally net exporter of:		total agricultural Land (1994)		63%
Forested	35%	live animals, fish and fish products and cereals		Share of private sector in total agricultural output (1992)		58%

ISSUE	STATUS OF REFORMS	OBJECTIVES PROPOSED ACTIONS
1. Macro-Economic Framework for Agriculture A. Prices/Subsidies	**Significant but incomplete liberalization of markets prior to independence, little further progress since independence** • All controls on producer and consumer prices removed prior to independence • Floor prices (above parity) for wheat, oilseeds, sugarbeet and tobacco; plus subsidies on milk, oilseeds, sugarbeet, cattle, olive oil, animal breeding, seeds and fertiliser • Large-scale purchase of wheat by State Directorate for Commodity Reserves, which distorts seasonal prices	**Distortion free marketing and incentive system and a more open trade policy** • Phase out floor prices and input and output subsidies • Limit activities of SDCR to the acquisition, management and disposal of a small strategic reserve, which does not distort domestic prices
B. Trade Policies	• WTO Accession in progress • Import quotas terminated mid 1996 and tariff structure simplified into a single ad valorem tariff of 5-25%. • System of variable levies being expanded and level of protection being increased. Is now the major source of protection. Erratic, ad hoc application of these variable levies has become a major source of price instability. • Few trade agreements. Currently negotiating multilateral agreements with the EU and CEFTA and bilateral agreements with Slovenia and Macedonia FYR	• Complete accession to WTO. • Drop variable levies and reduce and rationalise overall levels of protection • Additional multilateral and bilateral trade agreements
C. Taxation	• Low sales taxes on agricultural products (5%) and services (10%) with exemptions for agricultural machinery, veterinary services, fertiliser and cattle feed • Land tax abolished in 1996	• VAT to be introduced in 1997. Proposed level of 22% with no exemptions.

ISSUE	STATUS OF REFORMS	OBJECTIVES PROPOSED ACTIONS
2. Land Reform and Farm Restructuring	**Small-scale private farms predominated before independence, slow progress with privatization and re-structuring of the remaining ex-Social Sector Enterprises** • Small-scale private farms account for 63% of agricultural land but are very small (average 2.9 ha) and highly fragmented • Ex-Social Sector Enterprise land transferred to the State in 1991. Government not legally bound to privatize this land and has made little progress with leasing or sale to the private sector. • Conditions for an active land market not yet in place: property rights remain unclear due to delays in approving new land laws; and there are major inconsistencies between the land registry and the cadastre.	**Private ownership of all land; secure, transferrable property rights; and an active land market** • Enact laws on land consolidation and enforce existing inheritance laws to prevent further land fragmentation • Lease state land subject to ownership claims; sell all other state owned land • Passage of new law on Land Ownership and Property Rights • Passage of amendments to Cadastre and Land Registration Laws to make them compatible; update and reconcile Cadastre and Land Registry
3. Competitive Agro-Processing and Services for Agriculture	**Limited progress with privatization and demonopolization relative to pre-independence situation** • Approximately 50% of the equity in agricultural and agro-industrial enterprises is now in private hands • Slow progress with the privatization and restructuring (unbundling) of the larger remaining Kombinats and AgroKombinats due to their complexity and size, their overvaluation, and the operating procedures of the Croatian Privatization Fund • Seed and Animal Breeding Services still wholly under State Control • Removal of import quotas on fertiliser, tractors and seeds in 1996 but continuing high levels of protection for these and other farm inputs, which raises farm costs • Slow emergence of private sector input and output marketing structures and agencies to replace Kombinats and AgroKombinats • Agro-processing sector continues to receive very high levels of protection so raising consumer prices.	**Competitive, privately owned agroprocessing, input supply and service subsectors operating with minimal government protection** • Privatize and unbundle the remaining Kombinats and AgroKombinats • Register these companies and develop a secondary market for their shares • Passage of new legislation for seeds and animal breeding, and development of new institutional structures • Reduce protection on farm inputs • Encourage competition among domestic suppliers and from external sources • Reduce protection for agro-processors and encourage competition among domestic suppliers and from external suppliers

ISSUE	STATUS OF REFORMS	OBJECTIVES PROPOSED ACTIONS
4. Rural Financing	**Slow emergence of a viable private banking system which is active in the agriculture sector**	**Viable financial institutions serving the agricultural sector efficiently**
	• Massive program of subsidised credit financed by treasury was terminated in 1992	• Terminate credit activities of SDCR
	• Subsidized credit now restricted to very small amount of in-kind seasonal credit provided by State Directorate for Strategic Reserves	• Complete restructuring and rehabilitation of traditional commercial banks
	• Continued high interest rates and shortage of credit	• Continued support for improving bank supervision
	• Slow progress with privatization, rehabilitation and re-structuring of traditional commercial banks and the severance of their ownership links with the Kombinats and AgroKombinats	• Support the development of collateral instruments suited to agriculture and support training of agricultural lending officers • Guide interim use of the ADF and phase it out as soon as possible
	• New generation of smaller commercial banks beginning to emerge but with limited interest in lending to the agricultural sector	
	• MAF developing an Agricultural Development Fund (ADF) to provide subsidized credit to small-scale private farmers (approx $US 20m), as an interim measure until banking system becomes more active in the agricultural sector.	
5. Institutional Framework	**Slow adjustment of institutional structure to a role suited to a market-oriented economy**	**Efficient and effective public sector administration and support for commercial private agriculture**
	• Bulk of MAF budget is still allocated to direct subsidies for production and processing	• Reform existing system of incentives and the use of MAF budgetary resources
	• Some progress with the development of a public extension system but the institutional framework and links with the private sector are still weak	• Continued support for reform of extension program
	• Efforts to establish an Agricultural Research Council as the basis for focusing research and linking it with extension have made little progress	• Continued efforts to reform research system • Continued efforts to increase the institutional capacity for policy analysis
	• Limited MAF response to donor efforts to increase capacity for policy analysis	• Complete reform of legislation on seeds and animal breeding and develop new institutional structures.
	• Continued resistance to limiting government's role in seed and animal breeding to monitoring and supervision and allowing private participation	

ESTONIA

Total Population	1.6 mil	Agriculture in GDP 1996	4.8%	1995 agricultural output as percentage of 1986 levels	47%
Rural Population	28%	Food and agriculture in active labor (1995)	12.1%	1995 livestock sector as percentage of 1986 levels	52%
Total Area	4.52 mil ha.	Food and agriculture in export (1996)	16%		
Agriculture area:	1.46 mil ha.	Forestry share in exports (1995)	15.8%	Share of livestock (1996)	51%
		Food and Agriculture in imports (1996)	15.6%	Share of independent family farms in total cultivated land (1995)	29%
Arable land	77.5%	Traditionally net exporter of processed foods: eggs, butter, cheese and flax, in 1995 Estonia became a net importer.		Share of small subsidiary plots in total cultivated land (1995)	23%
Drained	66%				
Orchards	1.0%				
Natural Grassland	21.5%				
Forested	44.7%				

ISSUE	STATUS OF REFORMS	OBJECTIVE PROPOSED ACTIONS
1. Macro-economic Framework for Agriculture	**Completed price and market liberalization with almost no policy distortion.**	**Maintenance of an open market-oriented agriculture sector.**
A. Prices/Subsidies	• All prices are freely determined. • No government procurement. • No government price supports. • Modest deficiency payments were introduced for grain producers and dairy cattle in 1998. • A 25% grant is provided for investments on a competitive basis as of February 1998. • Currency has been successfully stabilized. • Competitive markets have been created based on minimal barriers to entry. • Budget financed rural credit fund providing credit with subsidized interest rates to agriculture and rural areas is still functioning.	• Continue commitment to liberalized prices and markets. • Avoid introduction of government intervention.
B. Trade Policies	• No export or import restrictions on agricultural trade. • No tariffs for exports and imports. • In compliance with IMF agreement. • Agreements with EU and EFTA countries.	• Remain a completely open economy and become a trading center for the Baltic states. • Avoid introduction of new protectionist measures.
C. Taxation	• Principle of equality is prevailing in the taxation policy. • Exemption from income tax for small private farmers.	• Do not create new distortion by introducing favorable taxation to certain groups.

ISSUE	STATUS OF REFORM	OBJECTIVE PROPOSED ACTIONS
2. <u>Land Reform and Farm Restructuring</u>	**Land is being restituted but the speed of reforms is curbed by financial, technical and legal impediments.** • Nearly all of the assets of state and collective farms have been privatized. • Majority of the land is still held in various larger incorporated farms which descend from the former collective and state farms. • Over 70% of agricultural land is still state owned. • Restitution claims have been settled only for about 25% of the agricultural area, 50% of the agricultural land has not been claimed. • The private farming structure has been developing towards smaller sized family farms with an average size of 24 ha. • Actual process of formal transfer of land titles to the newly established private farms has been slower than desirable. • The lack of security and transferability of leasehold has become a major impediment to the access of financial markets by those farmers that use unregistered lands. • Large proportion of the necessary farm drainage infrastructure is in poor condition. • Maximum land rental is set at 20% above the land tax, which implies an artificial upper boundary at 1.2% of the land value.	**Viable system of independently operated, privately owned farms.** • Accelerate land reform via increasing resources and coordinating activities like cadastre registration and restitution among involved institutions. • Introduce incentives to register land as soon as possible, e.g. tax breaks or other financial incentives. • Develop framework for a functioning land market especially by clarifying leasehold rights and obligations, establishing market based rental rates for leasehold properties, and facilitating the transactability of leasehold rights. • Include state land into this market and resolve the issue of how it will be disposed (either sale or long-term leases). • Upgrade and increase efficiency of farm drainage infrastructure. • Review present restrictions on foreign ownership and allow for more foreign participation. • Remove controls on land rents.
3. <u>Competitive Agroprocessing and Services for Agriculture.</u>	**Privatization of agroprocessing has nearly been completed.** • Companies are in the process of adapting to the market economy, but need significant levels of investment and are highly indebted. • Overcapacities in certain processing industries, e.g. dairy industry. • In some cases, excellent commercial institutions, often with the participation of foreign firms, have emerged.	**Internationally competitive agroprocessing and efficient services for agriculture.** • Remove preferential treatments and open agro-industry to foreign capital investments. • Foster the establishment of farmers' machine cooperatives. • Facilitate the upgrading of agroprocessing to meet European Union standards in all sectors still lagging behind.

ISSUE	STATUS OF REFORM	OBJECTIVE PROPOSED ACTIONS
4. Rural Financing	**Banking system has been largely privatized.** • Rural financial markets need to be further developed. • Modern banking system is developing. • Rural lending has originated almost entirely from budget financed credit fund at subsidized interest rates. • Lending to agriculture and rural sector has been low, but emphasis has been placed on increasing it. • Lack of an independent lending tradition based on risk assessment and risk taking by banks.	**Sustainable rural financial services.** • Train staff in financial sector to address the banking and financial requirements of the agriculture sector. • Train farmers in preparation of business plans and credit proposals. • Support the emergence of private banking structure. • Phase-out interest rate subsidies for rural lending. • Develop a rural guarantee fund to give partial guarantee on rural loans on a fee basis.
5. Institutional Framework	**Adjustment of the institutional framework is at a fairly advanced stage.** • Ministry of Agriculture has mainly regulatory functions. • Development of agricultural extension services will receive public funding from the agricultural loan. • Reorganized agricultural research and education is constrained by a shortage of funds and lacks market responsiveness. • Food Quality Control do not yet meet international requirements except in few sectors.	**Strategic framework for assistance to the rural sector.** • Accelerate technological transformation according to the needs of the private sector. • Set up project related research funding system, and demand oriented agricultural education. • Focus on rural poverty and regional development. • Improve public and develop private advisory services. • Establish monitoring and testing capacities for food quality to meet EU standards.

HUNGARY

Total Population	10.2 mil	Food and agriculture in GDP 1997	8%	Agricultural output in 1997 as percentage of 1990 level	76%
Rural Population	27%	Food and agriculture in active labor (1997)	12%	Livestock production in 1995 as percentage of 1990 level	66%
Total Area	9.3 mil ha	Food and agriculture in exports (1997)	22%	Share of livestock in agriculture (1995)	45%
Agriculture area:	6.2 mil ha	in imports (1997)	7%	Arable area in private use (1995)	98%
Arable land	70%	Traditionally net exporter: grain, meat, vegetable oil, processed and unprocessed fruits and vegetables, wine, livestock products and poultry.		Share of independent full and part-time family farms in total arable area (1997)	49%
Orchards	5%				
Irrigated	3%				
Forested	19%				

ISSUE	STATUS OF REFORMS	OBJECTIVES PROPOSED ACTIONS
1. Macro-economic Framework for Agriculture A. Prices/Subsidies	**Food and agriculture operates in a macroeconomic and trade environment with direct links to the world market; support programs and import tariffs, however, represent distortions.** • Agricultural producer and consumer prices were liberalized in 1990-1991. • Producer prices are about 30% below EU level. • Consumer prices are close to US domestic price level, but below EU levels. • Agriculture producer prices somewhat improved in 1995-97 relative to non-agricultural prices. • Modestly declining, but still significant subsidization of agriculture (about US $500 million, 1.1% of GDP in 1997). • Market support programs, (47% of total subsidies) mainly export subsidies, represent the most problematic component of support program: • Structure of support programs is changing in favor of direct support to producers such as investment grants and efficiency payments. • Minimum price programs exist but have not been effective so far.	**Liberal incentive and market system with minimal Government intervention.** • Create predictable and consistent system of various Government's policy instruments used in agriculture. • Revise existing support programs and continue the reduction of budgetary support in real terms. • Focus support programs on efficiency enhancement. • Avoid the use of minimal price programs and relate programs, if any, to world market prices rather than average cost of production.

ISSUE	STATUS OF REFORMS	OBJECTIVE PROPOSED ACTIONS
B. Trade Policies	• Tariffs for food and agriculture products are rather high (30-40%). • 8% additional export tariffs was introduced in March 1995, and it was phased out by mid-1997. • Licensing and quotas regulate the export of a few products including grain. • So called "Product Councils" incorporating produces, processors and traders work with the Government in implementing market intervention and trade policies. • Hungary became a full member of OECD in March 1996. • Hungary's non-compliance with the WTO agreement regarding export subsidies was a significant problem for trading partners within the WTO. • CEFTA is providing a framework for increased sub-regional agriculture trade. • Foreign trade is fully privatized and demonopolized.	• Implement WTO agreements on subsidies and tariff reduction. • Remove remaining export licensing and quotas. • Fully implement anti-monopoly legislation. • Pursue active trade policy to improve market access for Hungarian food and agriculture products.
C. Taxation	• By 1997 all producers marketing any agricultural products were to be registered at the tax authorities. • Government support programs are providing benefits, tax returns, nonregistered agricultural producers, do not receive any support. • 85% of fuel excise tax is refunded up to 50 liters/ha.	• Improve tax administration and tax collection in general. • Increase taxation of informal segments of agriculture while continuing to decrease taxation of formal sector as well as reported personal incomes. • Provide increased tax incentives for investment from properly reported corporate and personal incomes.

ISSUE	STATUS OF REFORMS	OBJECTIVE PROPOSED ACTIONS
2. <u>Land Reform and Farm Restructuring</u>	**The major initial tasks of land reform are close to completion.** • Land privatization included compensation of former owners and distribution of rest of collective land for members. • The compensation was completed by 1997, and 90% of land of new owners was physically identified. • Physical distribution of land for collective members is lagging somewhat behind. • First phase of collective farm reorganization was completed in 1995. Active members received 41.5%, previous owners 38.7%, former members 14.4% of assets. • The restructuring of new collective structures is continuing, the share of cooperative sector in arable land use was less than 30% in 1997. • Initial phase of state farm privatization was completed in 1996. 28 farms remained in majority state ownership (2% of land) out of the initial 121. • About 50,000 full-time family farms and about 1.2 million part-time farms cultivate 48.2% of arable land in 1997. • 18.7% of arable land is used by about 4000 incorporated private larger farms and by the 28 remaining state farms. • Titling of privatized land is lagging behind. In mid-1996, 55% of the beneficiaries of compensation and only 10% of land share owners received titles. On the whole about 20-25% of agriculture area still requires titling. • Land market is constrained by a three year moratorium on sale of land received through compensation, foreigners and legal persons can not own agriculture land there is a 300 ha upper limit set for individual land ownership.	**Privately owned smaller and larger viable farms as the dominant component of farming system with service and transferable ownership rights.** • Accelerate the processing of titling new privatized land. • Amend land ownership and land market regulations by: (a) allowing land ownership for legal persons (companies and corporations); (b) removal of upper limits for land ownership; and (c) allowing agricultural land ownership for foreign citizens. • Refine the legal framework for cooperatives in agriculture providing more transparency of ownership and framework for easy further restructuring and division of cooperative farms. • Introduce measures to facilitate a speedy consolidation of land ownership and changes in farm sizes. • Develop a strategy for further privatization of remaining state farms.

ISSUE	STATUS OF REFORMS	OBJECTIVE PROPOSED ACTIONS
3. Competitive Agroprocessing and Services for Agriculture	**Privatization of agroprocessing/input supply marketing and services is almost fully completed.** • Upstream and downstream sectors were privatized using tenders with entrepreneur to management and workers to obtain a limited percentage of share. • Foreign investors obtained 45% of ownership and made significant investments in 1994-96. • The upstream and downstream sectors are demonopolized and competitive. Input supply is integrated into the West European input supply system. A full selection of modern equipment and inputs are easily obtainable all over the country. • Gross output of agroprocessing in 1997 was around 95-96 % of pre-reform level.	**Competitive, privately owned agroprocessing and input supply.** • Fully implement anti-monopoly regulations. • Implement EU conform quality and safety standards for agricultural imports and exports. • Improve contract agreements and market transparency. • Promote research and development of new products and markets.
4. Rural Financing	**An appropriate financial system for privatized agriculture is not fully in place.** • Financing in agriculture is improving but still below the needs of a market based privatized agriculture. • High interest rates and the lack of collateral seriously limit lending to agriculture. • 30% interest rate subsidy is provided both for short term and investment credits. • Restructuring and privatization of banks with major portfolio in food and agriculture in process. • National system of rural credit cooperatives was created with EU. • Mortgage Bank was created in 1997 together with appropriate mortgage laws. • The use of warehouse receipts as collateral was introduced in 1997.	**Viable financial institutions serving the agricultural sector efficiently.** • Phase out credit subsidies. • Promote the emergence of competitive insurance services for agriculture. • Create a framework for use of warehouse receipts.

ISSUE	STATUS OF REFORMS	OBJECTIVE PROPOSED ACTIONS
5. <u>Institutional Framework</u>	**Institutional structure was reformed, quality public services, however, is not in place.** • Ministry of Agriculture operates in a market conform way. • Education system has been mainly adjusted to emerging new conditions. • Reorganization of the research system is planned for the near future. • Information system required by a market based agriculture is only partially in place. • Public activities (government research-education) in agriculture are seriously hampered by budgetary difficulties. • Agricultural extension system was created, including a system of township advisors and private advisory services (subsidy is provided).	**Promote the use of and support for commercial and private agriculture.** • Complete the reorganization and improve quality of public agricultural administration to the needs of a market economy and forthcoming EU accession. • Complete the reform of agricultural education and research.

LATVIA

Total Population	2.48 mil	Food and agriculture in		1996 agriculture production as a	
Rural Population	31%	GDP (1996)	9%	percentage of 1990 levels	42%
		Food and agriculture in		1996 livestock production as a	
Total Area	6.5 mil ha	active labor (1996)	18%	percentage of 1990 levels	34%
Agriculture area:	2.5 mil ha	Food and agriculture in		Share of livestock (1996)	45%
		export (1996)	16%	Agriculture area in private use (1996)	88%
Arable land	66.9%	in import (1996)	12%	Share of private sector in total	
Pastures &	32%	Traditionally net exporter		agricultural output (1996)	95%
Meadows	62%	of livestock products: meat,			
Drained	43%	milk, and eggs			
Forested	44%				

ISSUE	STATUS OF REFORMS	OBJECTIVES PROPOSED ACTIONS
1. Macro-economic Framework for Agriculture	**Almost fully completed liberalization, but with relatively high import duties.**	**Distortion free, efficient internationally competitive agricultural sector.**
A. Prices/Subsidies	• Prices are freely determined. • No Government procurement quotas. • Minimum guaranteed prices for a small quantity of grain set by the Government for state reserves close to current market prices. • No deficiency payment systems. • Direct payment supports (Lat 4 million) for high grade breeds and seeds, and flax in 1996. • Pressure to introduce EU type protective measures (Draft law on agriculture).	• Keep commitment to liberalized prices. • Foster the development of price information systems for transparent and timely access to cost changes by all concerned. • Do not introduce new credit subsidization or high price guarantees. • Phase out also all implicit subsidies and avoid new subsidization of agriculture.
B. Trade Policies	• Export of agricultural products is deregulated. • Import regime has become more restrictive Variable tariff on agricultural import commodities on average 40% ad valorum (import tariffs for butter and cheese as high as 55%). • Not in compliance with IMF agreement except budget deficit ceiling limit. • Grain import/export requires licence. • 3% tax on grain trading transactions except those of primary producers. • Temporary export subsidies for dairy products in 1995. • Export taxes on unprocessed logs and animal hides.	• Maintain an open economy policy to stay competitive. • Rationalize tariffs on low uniform ad valorum levels and avoid frequent changes to reduce uncertainty and corruption. • Uphold IMF tariff, export restriction and import quota agreements. • Promote emergence of wholesale marketing by private traders by providing facilities, and allow new free entry to or exit from the sector. • Proceed with demonopolization, corporatization, and phased privatization of trading enterprises. • Strengthen legal framework to improve binding of transactions.

ISSUE	STATUS OF REFORM	OBJECTIVES PROPOSED ACTIONS
C. Taxation	• Tax burden on farmers is relatively low; tax exemptions are being phased out (no property tax on agricultural land, and preference in social security taxes).	• Reduce differentiation in tax rates to minimize distortive effects on resource allocation through unequal burden. • Introduce a social security system tailored for the conditions of private farmers.
2. Land Reform and Farm Restructuring	**Government committed to transforming agriculture into an efficient and dynamic sector by encouraging the development of a market-based, predominantly privately owned production system, but considerable further adjustment is needed.** • Significant progress in the privatization of agricultural land. • Land restitution is at an advanced stage. • Newly formed large agricultural joint stock companies are inherently unstable, because individual shareholders have the right to transform their shares into physical assets any time they like. • About 75,000 private farms have been restituted. Only a few claims have yet to be processed. • Registration/titling is advancing slowly, though intensive TA being provided. • About 65,000 private properties have been registered in cadastre registration, but full titling is lagging, and about 43,000 have been registered in the land book. • Clear formal mechanism for valuation, buying, and selling of land is missing and prevent the development of fully functioning land markets. • Drainage systems do not meet the needs of smaller privatized farms and require rehabilitation or comprehensive repair and maintenance, also to curb negative environmental impact.	**A farming system based mainly on private ownership of land and a working land market for efficient and fair asset transfer.** • Accelerate land survey, registration and titling to foster the emergence of a land market. • Establish a consistent land valuation system and information flow and a formal mechanism for land transactions. • Make leasing rights freely tradeable. • Complete restitution. • Help organize local and regional farmers' cooperatives. • Rehabilitate and modernize key rural infrastructure, including access roads, electricity, and water to new private farms. • Repair and redesign drainage systems to meet needs of new farming structure.

ISSUE	STATUS OF REFORM	OBJECTIVES PROPOSED ACTIONS
3. Competitive Agroprocessing and Services for Agriculture.	**Good progress in privatization of agro-industries.** • Almost the whole large-scale state and collective agroprocessing industry is privatized. • Small-scale businesses have started to emerge. • The agro-industries have suffered reduction in capacity utilization by 30-50% and lack modern equipment and design. • Development of the quality of both existing and new products through innovation is limited. • Quality control and measuring systems are inaccurate or not working at all, limiting export opportunities. • Development of marketing infrastructure in the food chain stays behind the needs of the large number of smaller farms and enterprises. • Anti-monopoly legislation exists, but not strictly enforced	**Efficient, privately owned agrobusiness firms subject to market forces, and agroprocessing industries with high quality products which can compete in world markets.** • Complete the privatization of agroprocessing industries including the fully state owned enterprises. • Continue a policy that effectively stimulates competition. • Design and implement a program to encourage rural SME development • Support for restructuring and capitalization of enterprises based on their competitive merits. • Create an environment favorable for foreign investment. • Support development of a commercial network of private input and output markets. • Support the introduction of product quality control and inspection systems that meet EU or US standards as soon as possible to gain access to these markets. • Facilitate farmer/processor/consumer market information systems.
4. Rural Financing	**A private banking system is emerging with mixed results. Strengthened bank supervision capacities in the Central Bank are needed.** • Only limited availability of medium/long-term rural credit to the agricultural sector through commercial banks. • Credit is available for the agricultural sector, particularly for the newly emerging private farmers, but lenders lack knowledge about agricultural loan handling.	**Viable financial institutions serving the agricultural and rural sector efficiently.** • Implement generally accepted accounting and external audit principles. • Strengthen prudent banking regulation and supervision. • Train loan officers in agricultural lending including credit risk assessment and portfolio management. • Assist new private farmers in start-up period with provision of credit, training, and services.

ISSUE	STATUS OF REFORM	OBJECTIVES PROPOSED ACTIONS
5. Institutional Framework	**Emerging large number of private farms face insufficient support services, including extension and research.** • Government structure has not been adjusted to the needs of reformed agriculture • Newly created private farms have only little farming and farm management skills; technical assistance is needed to develop institutional and human capacity to support emerging private entrepreneurs. • Latvian Agricultural Advisory Service (LAAS) was established in 1992 and has been receiving substantial TA from bilateral and multilateral sources. • Agricultural research and education lacks incentives for coordination and problem orientation, research topics, curricular need to be updated for applicability.	**Efficient and effective public sector administration and support for private agriculture.** • Reorganize the Ministry of Agriculture and regional agriculture administration according to the needs of a market economy. • Encourage and strengthen research system to focus on applied research, transfer and adoption of foreign technology, and the design of cost effective extension activities. • Continue to develop independent effective and far reaching sector farm management and technical advisory services as well as agricultural extension services suitable to all possible sizes of private farms. • Reorganize and strengthen linkages between research, teaching, and extension to increase effectiveness and efficiency. • Strengthen the capacity for market-oriented policy analysis.

LITHUANIA

Total Population	3.7 mil	Food and agriculture in GDP		Agricultural output in 1996 as percentage of	
Rural Population	32%	1996	10%	1989 levels:	58%
		Food and agriculture in active		Crop production	72%
Total Area	6.5 mil ha	labor force (1996)	24%	Livestock production	44%
Agriculture area:	3.5 mil ha	Food and agriculture in		Share of livestock in agriculture (1996)	45%
Arable land	84%	export (1996)	17%	Agricultural area in private use (1996)	88%
Orchards	1.7%	in import (1996)	13%	Share of family private farms in total	
Pasture	14.3%	Traditionally net exporter of		agricultural output (1996)	75%
Forested	32%	livestock and dairy products			

ISSUE	STATUS OF REFORMS	OBJECTIVES PROPOSED ACTIONS
1. Macro-economic Framework for Agriculture A. Prices/Subsidies	**Significant but incomplete liberalization of agricultural markets:** • Major reform of price support systems in progress, reducing the number of products covered and focusing on higher quality products. • Quotas for which minimum prices were paid under the minimum marginal price program were drastically reduced. • A grant system to improve productivity is becoming the major component of the agricultural support program. • Agricultural budget for agricultural subsidies was reduced. Reduction in budget for agricultural subsidies reached 16.2% in real terms.	**Distortion free, efficient and internationally competitive agricultural sector:** • Continue liberalization of prices and discontinue all government purchase for domestic market regulation purposes • Develop an agricultural policy strategy for shifting agricultural support away from direct subsidy and price control toward a focus on efficiency enhancement.
B. Trade Policies	• Limited progress in re-orienting the country's agricultural exports to non-FSU. • Tariff on agricultural commodities ranging from 10%-30%. • A temporary export subsidy for beef and dairy has been approved in early 1997. • Remaining export bans were converted into export tariffs of not more than 60% with the exception of sugar. • Product markets are constrained, particularly at peak harvest time, due to lack of storage facilities (e.g. perishable goods).	• Remove all quantitative (non-tariff) trade restrictions (including temporary export ban on wheat). • Rationalize tariffs on low uniform levels and avoid frequent changes to reduce uncertainty and corruption, and uphold IMF tariff, export restriction and import quota agreements. • Improve legal framework supporting trade, especially the domestic and international payment system. • Promote emergence of wholesale marketing by private traders by privatizing remaining state facilities.
C. Taxation	• Until 1997 farmers have not had to pay land ownership tax or income tax. • Fuel excise tax is partially refunded to agriculture producers. • A new tax system for private agriculture will be introduced in 1997.	• Develop a transparent agricultural taxation without discrimination.

ISSUE	STATUS OF REFORMS	OBJECTIVES PROPOSED ACTIONS
2. Land Reform and Farm Restructuring	Government committed to transforming agriculture into an efficient and dynamic sector by encouraging the development of a market-based, predominantly privately owned production system, but the progress to date has been mixed. • Significant progress in the privatization of agricultural land. • Restitution began but was suspended in early 1997 when the laws were changed to broaden and improve the procedures. Restitution resumed in July 1997. • The removal of the moratorium on land ownership by local legal entities is being considered. • Special privileges for agricultural companies (i.e. large-scale farms) to lease land for another 5 years expired at the end of 1996. • In 1997 there were about 196,000 private farms, holding about 42% of available agricultural land, 342,700 household plots accounting for 24.5% of agricultural land. • A significant proportion of arable land was left idle due to lack of inputs and working capital, uncertainty over property rights, and poor drainage conditions. • Clear legislation on land registration and transactions are missing and delay development of fully functioning land markets.	A farming system based mainly on private ownership of land and a working land market for efficient and fair asset transfer. • Speed up land survey registration and titling service as a priority to developing a fully functional land market. • Take steps to shorten the registration time for associations and cooperatives of private farmers by reviewing registration procedures. • Establish a framework for consistent land valuation and for the dissemination of land market information. • Improve the design of the legal and institutional framework that provides for financially sound mortgage operations. • Formalize clear rights to own and sell land and allow sales of land. • Improve tradability of leasing rights by increasing efforts to ensure land markets function. • Promote new roles for western type cooperatives in agriculture. • Provide support for the further transformation of agricultural companies ('bendroves').

ISSUE	STATUS OF REFORMS	OBJECTIVES PROPOSED ACTIONS
3. Competitive Agroprocessing and Services for Agriculture	Considerable progress in legal reforms, but lack of enforcement and limited progress in privatization of large-scale agro-industries. • First phase of privatizing agroprocessing is nearly complete. By the end of 1997 only 8% of shares will be owned by the state, farmers will own 44% and employees 48%. • A draft law to make farmers' shares in food processing companies tradable under the same conditions as shares in other companies is being considered. • Foreign participation in marketing agroprocessing is minimal. • The agro-industries are over-sized and inefficient with outdated technology and equipment. • The increasing emergence of private entrepreneurs is bringing about an increased level of competition as a result. • Development of the quality of both existing and new products through innovation is severely curtailed, thus reducing export opportunities. • Bankruptcy laws have been enacted and the special treatment clause which continued to protect agroprocessing industries was phased out in July 1994. • The problem of delayed payments to farmers is still widespread, although laws were enacted in 1994 which require agroprocessing enterprises to pay farmers within 15-30 days. • Development of information system and marketing infrastructure in the food chain is largely behind the needs of the large number of smaller farms and enterprises.	Efficient, privately owned agrobusiness firms subject to market forces, and agroprocessing industries with high quality products that can compete in world markets. • Develop a policy that effectively stimulates competition policy and actively removes monopoly distortions. • Sell remaining state shares of agroprocessing. • Create transparent ownership in agroprocessing enterprises where primary producers own more than 50% of the shares. • Foster the start-up of new commercial ventures based upon new agricultural and agroprocessing technology. • Create farmer/processor market information systems. • Modify transport, storage, and communications facilities. • Expand and upgrade public facilities for output marketing (farmers' markets, etc.).

ISSUE	STATUS OF REFORMS	OBJECTIVES PROPOSED ACTIONS
4. Rural Financing	**Creation and development of a sound financial system is under way, but is greatly interfered and hampered by ad-hoc Government intervention and a lack of long-term capital and knowledge about loan handling.** • Sufficient credit is available for corporate farms, while only limited credit is available for the bulk of private farmers. • Almost complete absence of medium/long-term rural credit to the agricultural sector that suffers from a severe lack of productive assets and capital, and the full range of inputs. • Significant progress in the privatization of banking institutions. • Rural Credit Guarantee Fund was created in 1997. • Subsidized credit from the budget is reduced to short-term seasonal credits in 1997.	**Viable financial institutions serving the agricultural and rural sector efficiently.** • Implement generally accepted accounting principles. • Strengthen prudent banking regulation and supervision. • Train loan officers in agricultural lending including credit risk assessment. • Phase-out remaining subsidized credit and eliminate ad-hoc injection of credit from the budget. • Facilitate the emergence of rural credit cooperatives.
5. Institutional Framework	**Emerging large number of private farms face insufficient support services, including extension and research.** • Although the Lithuanian Agricultural Advisory Service (LABS) is developing well, some aspects of the farm extension system remain underdeveloped, and there is a need to increase the flow of information from research and education organizations to the advisory service and farmers. • Research direction of state research institutes and universities is only slowly changing from the old axioms in agriculture towards support for the new emerging farm systems and products. • Institutional strengthening of the Ministry of Agriculture and Forestry and downsizing of the Ministry was implemented in early 1997.	**Efficient and effective public sector administration and support for private agriculture.** • Reorganize and strengthen problem and long-term prospect focused research system. • Further develop a private sector farm management and technical advisory service as well as agricultural extension services. • Continue to emphasize ecologically sound production systems.

MACEDONIA, FYR

Total Population	2.2 million	Food and agriculture as percentage of 1996 net material product	15%	Agricultural output in 1996 as percentage of 1989 level	90%
Rural Population	45%	Food and agriculture in active labor (1996)		Livestock production in 1996 as percentage of 1989 level	80%
Total Area	2.0 million ha.	Food and agriculture in exports (1996)	9%	Share of livestock in agriculture (1996)	35%
Agriculture area:	1.2 million ha.	in imports (1996)	19%	Private agricultural area (1996)	80%
Arable land	44%	Traditionally net importer of grain, sugar, vegetable oil and livestock products.	30%	Share of independent private farms in total arable area (1996)	80%
Orchards/Vineyards	15%				
Irrigated	18%			Share of private sector in total agricultural output (1996)	65%
Forested	35%				

ISSUE	STATUS OF REFORMS	OBJECTIVES PROPOSED ACTIONS
1. Macro-economic Framework for Agriculture A. Prices/Subsidies	**A standing agreement with the Fund and Bank exists on a viable, medium-term macroeconomic framework** • Government-financed price premiums for wheat, sugarbeet, sunflower, and other oilseeds and milk will not exceed 1995 levels. • Fixed retail price of a standard loaf of bread changed to a price cap. • Retail prices for basic foods have been freed, except for the standard loaf of bread. • Fertilizer subsidy and seed rebate were eliminated. • Guaranteed prices eliminated for sugarbeet and sunflower. Government is no longer under obligation to award price premiums for sugarbeet, sunflower, other oilseeds and milk, and such payments ceased in May 1997.	**Maintain agreed macroeconomic framework** • Maintain a formula linking the guaranteed base prices for wheat and tobacco to world market prices so that such prices shall not exceed 100 percent of the world market prices in 1997 and 70 percent thereafter. • Remove the price cap on the standard loaf of bread when the anti-monopoly law is in place. • Eliminate credit rebates for agricultural production.
B. Trade Policies	• Customs harmonized tariff in place except those established by international treaty or for goods for humanitarian purposes.	• Remove remaining import quotas except those justified on security grounds. • Implement improvements in duty draw back scheme. • Constrain export subsidies to 1997 level and remove remaining export quotas except those justified on security grounds. • Establish in a trade policy unit that would monitor the country's trade regime, assess any proposals for change, and act as a focal point for the country's accession to membership in the WTO and EU.
C. Taxation	• Implicit taxation of primary products is being eliminated through fixed prices. • Farmers are exempt from income tax and there is no land tax.	

ISSUE	STATUS OF REFORMS	OBJECTIVES PROPOSED ACTIONS
2. Land Reform and Farm Restructuring	**Land is mostly under private ownership, but markets are not functioning.** • 80 percent of the arable land is under secure ownership by private farmers. • Draft law on land use (11/98) incomplete • Land markets are "sticky," i.e., there are high friction costs involved with the sale, rent or lease of land. • The law for privatizing the agri-kombinats has been passed but not yet implemented. • Private farmers on the average are small. However, in order to take advantage of economies of scale many farmers "informally" aggregate land in to larger operational holdings. This practice was not encouraged in the past and is not legally recognized.	**Fully functioning land market and privatized farms.** • Develop a law on land use that does not allow for Government dictation of cropping choices by farmers. • Complete the privatization of the Agri-kombinats (20% arable land). • Improve the operations of the land registration system to facilitate land sales.
3. Competitive Agroprocessing and Services for Agriculture.	**Agricultural chemicals and vegetable seed largely supplied by the private sector.** • Agricultural implement dealers limited. • Agroprocessing mixed with the dairy industry rapidly privatizing, but cold storage, canning and wine making still largely dominated by the agri-kombinats. • Private, refrigerated trucking companies rapidly becoming more important. • The supply of the seed of field crops dominated by Government Institutions and/or agri-kombinats.	**Greater development of private sector input supplies.** • Pass enabling legislation that encourages a broadening and deepening of the private sector seed industry.

ISSUE	STATUS OF REFORMS	OBJECTIVES PROPOSED ACTIONS
4. Rural Financing	**No formal rural financial intermediation services exist and rural residents have difficulty in accessing bank credit due to lack of collateral. Because of past experience, rural residents distrust the existing formal banking system. There is evidence of relatively high rural savings rates.** • Given the level of uncertainty, rural residents prefer to keep debt levels low or none existent and will self-finance new investments such as barns, cattle, production supplies, etc. There is no evidence, except among the agri-kombinats or their successors, of constraint on credit funds. • The biggest need for investment and working capital important for the agricultural sector lies with the upstream, e.g. input suppliers, machinery sales, etc., and the downstream, e.g., marketing and processing entities, etc.	**Development of financial intermediation services that inspire confidence.** • The development of savings and loan associations, or savings cooperatives, should be pursued. • Given the long lead time necessary to develop a strong financial system, implementation of an agricultural inputs guarantee program would support input suppliers, for example, in tapping outside capital sources to enable supply dealers to extend suppliers credit to producers.
5. Institutional Framework	**Ministry of Agriculture retains a semblance of ministry of production from the centralized system.** • The agricultural research system is starting to reorient following breakup of Yugoslavia system. • The extension system is adapting to demand driven system with enactment extension law 1/98. • The agricultural university system is largely out of date and is out of touch with the requirements for a market economy. • New Veterinary Law (11/97) encourages development of private practices; strong interest being shown in most areas.	**Reorientation of the Ministry of Agriculture from one of intervention to one of support for private, market driven agriculture production, including the provision of timely market information, adequate analysis of the implications of public policy choices in the agricultural sector.** • Restructure the agricultural research system so that it is relevant and sustainable for a small country that needs to be flexible in responding to new market opportunities, and opportunities presented by new, technological innovations and/or dramatic shifts in real or relative prices. • Modernize the course offerings and course contents of the Agricultural Faculty so that they are relevant and responsive to the needs of a market economy, particularly with respect to economics and marketing.

POLAND

Total Population	38.6 mil	Food and agriculture in GDP 1995	14%	Output Ratios 1993 / 1989:	
Rural Population	38%	agriculture alone	6%	total agriculture	85%
		agriculture alone in 1989	13%	crops	90%
Total Area	31.3 mil ha	Food and agriculture in active labor		livestock	79%
Agriculture area:	18.7 mil ha	(1995)	31%	Share of livestock in	
		agriculture alone	26%	agriculture (1993)	42%
Arable land	77%	Food and agriculture in exports (1995)	9%		
Orchards	2%	of which to EU	55%	Avg. per cap. real farmer	
Meadows &		of which to FSU	31%	income '93/'88	40%
Pastures	21%	Food and agriculture in imports (1995)		Inter-sectoral terms of trade	
Forested area:	27%	of which from EU	10%	(output/input prices) 1995	48
		Sales food proc./all industrial sales	49%	[1986 = 100]	
		(1995)	20%		

ISSUE	STATUS OF REFORMS	OBJECTIVES PROPOSED ACTIONS
1. Macro-economic Framework for Agriculture A. Prices/Subsidies	Market liberalization is advanced, although ad-hoc protection and support are obtained periodically by selected products, companies and sub-sectors. • Support for agriculture is in 4 categories: (i) direct subsidies to the farmers' social security system, (ii) price support and intervention purchasing, (iii) input subsidies and (iv) subsidies for farm modernization and rural infrastructure. In addition, state agencies provide preferential credit and credit guarantees on commercial loans. Total budget transfers have decreased in real terms by about 1.5% yearly in the last 4 years, excluding the off-budget support by the state agencies. The total budget for agriculture represents about 2.5% of GDP. • Support prices are defended by the state marketing agency (ARR - Polish acronym) for: bread wheat, rye, butter, skim milk powder, beef and pork carcasses, sugar and potato starch. • PSEs, in 1994, were 21%, almost totally attributable to price support (CSEs were -18%). • There is overlap between market intervention and government purchases of national food reserves, decreasing the transparency of support.	Limited intervention; a rule based safety net for catastrophic occurrences only; reform of farmers' social security and the tax regime • Make price support rule based, linking it to an objective measure, such as border prices and with sufficient average difference between intervention and market prices to allow the market to operate. Limit price support to key commodities. • Reform farmers' pension system which now costs over 70% of the total budget of the Ministry of Agriculture and Food Economy (MAFE), while contributions cover less than 10% of costs. • Discontinue distortionary and opaque support given by state agencies. • Restructure the Agricultural Marketing Agency (ARR - Polish acronym) into an implementation office for CAP directives; phase out the national food security stocking program.

ISSUE	STATUS OF REFORM	OBJECTIVES PROPOSED ACTIONS
B. Trade Policies	• As signatory to the WTO agreements, Poland has abandoned quantitative import restrictions and variable levies, but is allowed tariff protections similar to those of most WTO countries. The average weighted tariff on agricultural and food products is almost 24%, up from almost 20% prior to the WTO agreements. • A 10% duty on grain imports was reinstated by government decision of 1/28/97. A 20% duty was suspended in April 1996. The tariff does not apply to durum wheat, maize and soybeans. • The sugar support regime is copied from the EU, with WTO allowed subsidized export quotas (about 120,000 tons in 96/97, less than half the previous year's). Export subsidy is paid from domestic consumer levy.	• Resist demands for protection; interpret what is allowed under WTO conservatively and insist on quantified economic justification for all tariffs proposed; incorporate estimates of real exchange rate in evaluations. • Do not raise protection to EU levels, in spite of expected imminent accession, since (i) it is a (downward) moving target, (ii) it would slow down needed structural changes and (iii) the fiscal cost would be prohibitive. • Evaluate the long-term prospects for the sugar sector, assuming EU accession; make a plan for its rationalization based on full privatization and phasing out of export subsidies.
C. Taxation	• No income tax on revenue from most agricultural operations. • A land tax is levied on agricultural holdings as a function of soil type; selected investments can create a tax deduction • Rural cooperatives are taxed as corporations, this perceived as double taxation [corporate income tax plus personal income tax on dividend or profit distribution] by members. • 3% duty on all imports, to be lifted in 1997. • Tariffs on agricultural product imports are being reduced 10%/year, as of July 1, 1995. • No VAT on farm inputs and equipment; but a proposal to introduce a 7% VAT tax in 1997. • Zero VAT rate on agricultural products to be maintained at least till 1998, to allow for creation of VAT collecting rural institutions.	• Agriculture derived income above a threshold should be taxed. • Corporate tax on cooperatives should be abolished; profit distribution should be obligatory, with limited retained earnings allowed [US sub-chapter S corporate structure could be a guide] • Introduce VAT on inputs and products so as to comply with EU tax rules. • Levy agricultural tax only on land actually under production, so as to create incentive for voluntary set aside. • Introduce a pollution tax based on mineral book keeping to discourage misuse of fertilizers.

ISSUE	STATUS OF REFORM	OBJECTIVES PROPOSED ACTIONS
2. Land Reform and Farm Restructuring	**Production on the 18% of arable land that used to be in State farms has been privatized, but ownership of that land is less than 10% private. The average size of private farms, on the remaining 82% of arable land, is only 6 ha, in an average of more than 5 parcels.** • All assets and liabilities of former state farms were transferred, in Jan. 1992, into a state owned, off-budget, holding company, APA. Privatization of land and other assets, principally about 1,000 agro-processing enterprises, has been slow; debts are not being serviced. • In addition to the assets of the former state farms, APA also received custody of the Land Fund, consisting of a large number of, usually small, parcels of land, handed over to the state as a condition of eligibility for a full state pension. This system of "a pension for land" remains in existence. [retirees are allowed 1 ha.] • About 70% of arable land in APA's holdings is leased, often to small associations of individuals that used to manage the state farm; most of the rest is fallow. • Privatization of land is hindered by: reluctance to break-up large farms, pre-nationalization claims, political opposition against privatization and associated land sales to foreigners, and by vested interests of APA staff. • Land holding, and production, structure on the 80% of agricultural area in private hands is archaic, reminiscent of the situation in most of Western Europe immediately after the second world war. Land consolidation is less than 20,000 ha annually.	**Complete privatization of former state farms; accelerated land consolidation and development of land market.** • Speed up and complete the privatization of ownership of former state farm land; consider auctioning off the rights to privatization, within prescribed guidelines, to private land development companies operating in the local land market. Develop a sunset policy for APA and amend its by-laws accordingly. • Improve the operation of the land market by (i) fomenting the creation of a real estate brokers association, (ii) assist with the development of information on land transactions and on offers and requests for land and (iii) improve the agility of registration of land transactions. • Separate all aspects of land taxation from those pertaining to the land market. • Develop a potentially effective program of land consolidation, fully integrated with the land market. • Integrate the lands obtained from aspiring retirees into the land market, directly and through the consolidation programs.

ISSUE	STATUS OF REFORM	OBJECTIVES PROPOSED ACTIONS
3. Competitive Agribusiness System.	**Delays in privatizing agroprocessing; emergence of new private sector processing; large border trading companies have emerged from the human and goodwill assets of former state trading monopolies; agriculture service sector, notably marketing, remains weak.** • Privatization remains to be completed in the cereals, sugar, fruit and vegetables and meat processing subsectors. • Delays in privatization have maintained high cost enterprises, creating opportunities for new private entrants to exploit attractive margins. This, though, puts a question mark on the industry's competitiveness. • Standards and quality control remain to be adjusted to the requirements of a competitive market, with emphasis on the EU's "acquis communautaire." Industry interest associations that should undertake some of that work and lobby government for the rest are still embryonic. • Apparently large and well endowed and connected trading companies may pose a risk of monopolization of certain trades. The same companies are increasing their holdings of agribusiness assets, often as successors to state ownership.	**Complete privatization and harmonization of rules, regulations, standards and controls with the EU and support the development of interest associations and market instruments** • Develop a divestiture plan, with target deadlines, for all remaining state owned agro-industry enterprises, including privatization, bankruptcy and sales of assets. • Develop a timebound implementation plan for the part of the EU's "acquis communautaire" that pertains to the food processing industry, including the creation and adaptation of institutions and associated training and staff development. • Assist with the creation of wholesale markets, warehouse receipts and commodity futures contracts. • Support the development of interest associations in agribusiness activities. • Study the likely competitiveness of Poland's agro-processing sector within an enlarged EU.
4. Rural Finance	**The dominant institution in rural finance, the Bank for Food economy (BGZ) remains in state hands; interest rate subsidies are still used; private banking for rural finance remains highly selective.** • The former rural finance monopoly, BGZ, still has a large portfolio of non-performing assets belonging to former or current state owned enterprises. Repeated recapitalizations have absorbed the equivalent of $1.2 billion in budgetary resources. • A new cooperative banking law, introduced in June of 1994, effectively cements BGZ's position as the apex of a three tier cooperative banking system, with about 1,200 rural cooperative banks. A rival system based on about 300 rural cooperative banks struggles in the absence of official support. • The limited rural financial intermediation leads to a preponderance of cash transactions and savings that are held in cash or in physical assets.	**Viable rural finance system, served by agile financial institutions, incorporating the traditional rural cooperative banks.** • Resolve the core structural weaknesses of BGZ, both operationally and in its portfolio; privatize the bank. • Discontinue interest rate subsidies and the special government guarantees for BGZ deposits. • Create ways and means to give rural municipalities access to the capital market. • Develop a system that will allow disbursements of expected future EU structural funds through rural financial institutions. • Discontinue the credit guarantee operations of ARR, ARMA and APA, in favor of one transparent program operated by an accredited financial institution.

ISSUE	STATUS OF REFORM	OBJECTIVES PROPOSED ACTIONS
5. Institutions	**Institutional restructuring has been substantial and meaningful; selected adaptations to EU requirements are now needed.** • The Ministry of Agriculture and Food Economy (MAFE) was restructured under the MAFE and lately further streamlined as part of overall restructuring of government. The Minister is at the same time one of three deputy premiers. • In hindsight it may be questioned whether sunset policies for APA, ARMA and ARR, included in their acts of establishment, would not have been preferable over the current open ended existence. • The extension service was essentially recreated with a dual emphasis, on technical matters and on business matters. • Important institutions remain dependent on, often foreign, off-budget resources for their existence. This keeps them sometimes de-facto out of the mainstream of agriculture policy making and development. • The MAFE has taken an important initiative in 1996, to restructure the curricula of the agricultural high schools so as to make it more relevant to contemporary conditions. • The budget squeeze on institutes of higher learning has forced its management into entrepreneurial activities, bringing more "real world" focus into that part of academe. • The agricultural research establishment, notably government's 42 breeding farms, maintains institutions and practices that appear archaic but that may still be appropriate given the archaic nature of production and landholding of the vast majority of Polish farms.	**Modification of institutions dealing with the EU and changes in practices and methods of operation in step with structural changes in production and landholding.** • Continue and strengthen the on-going program of institutional adaptation to the requirements of the "acquis communautaire." • Bring all institutions under the purview of government by making them "on-budget" and have foreign support for them flow through the budget. • Make personnel remuneration in selected institutions market based, rather than being determined by general civil servant rules, so as to attract and keep "the best and the brightest." This is of particular importance in view of the upcoming negotiations for EU membership. • Complete the reform of agricultural education and associated research. • Study the expected future direction of agricultural research in Poland, both private and public, in support of the development of a research strategy.

ROMANIA

Total Population	22.68 mil	Food and agriculture		Gross Agricultural Output (GAO) in 1995 as	
Rural Population	45.1%	in GDP 1994	28.8 %	percentage of 1990 level	100.8%
Total Area	23.8 mil ha	Food and agriculture		Livestock production in 1995 as	
Agriculture area:	14.8 mil ha	in active labor (1995)	34.0 %	percentage of 1990 level	92.3%
Arable land	63.1%	Food and agriculture		Share of livestock in agriculture as	
Orchards, vineyards	3.9 %	in exports (1996)	8.7 %	% of GAO (1995)	40.4%
with Irrigation Facilities	13 %	in imports (199)	7.4 %	Arable area in private use (1996)	73.0%
w/o Irrigation	1.8%	Traditionally net exporter: live cattle and		Share of independent full and part-time family	
Forested	28.1 %	sheep, meat (pork, beef, poultry), grains, sunflower oil, wine		farms in total agricultural area (1996) 54.4%	

ISSUE	STATUS OF REFORMS	OBJECTIVES PROPOSED ACTIONS
1. Macro-economic Framework for Agriculture	Food and agriculture operates in a macroeconomic and trade environment with direct links to the world market. Lack of private market institutions, price transparency, high transport costs and poor logistics inhibit market development.	Liberal incentive and market system with minimal Government intervention. Support for the development of private market organizations, market information systems and better infrastructure and logistics.
A. Prices/Subsidies	• Gradual price liberalization started Oct-90. Agricultural producer and consumer prices were liberalized Feb-97. No minimum price schemes. • Producer prices are around their border parity, and below EU levels. Large share of the production is not marketed. Consumer prices are at their export parity, or slightly above in the case of importables, but below EU levels. • Regional price variation is significant, due to high transport costs, poor logistic and arbitrage. Absence of market information system contributes to regionalization of domestic trade in food products. • Intention to institute a minimum price scheme for wheat starting 1997/98. • Subsidies for agriculture reduced in 1997. However, they still account 2% of GDP (about 4000 bln. lei, equivalent of USD 570 million). One third of the subsidies have been used to pay contingent liabilities for 1995/6 harvest (interest rate subsidies, arrears for producer subsidies like premiums).	• Create predictable and consistent system of various Government policy instruments used in agriculture. • Revise existing support programs and continue the reduction of budgetary support in real terms. • Focus support programs on efficiency enhancement. • Avoid the use of minimal price programs and related programs, if any, to world market prices rather than average cost of production. • Develop and support initiatives for market information system (price and output).

ISSUE	STATUS OF REFORMS	OBJECTIVES PROPOSED ACTIONS
B. Trade Policies	• Tariffs for food and agriculture products reduced from a trade-weighted average of 80% to 27% starting Jun-97. • Licensing and quotas for exports and imports removed starting Jun-97. Preferential import quotas maintained in the framework of the bi- and multilateral trade agreements. • Romania joined CEFTA in 1997; the agreements with CEFTA, EU and Moldova are providing a framework for increased sub-regional agriculture trade. • Agricultural foreign trade is fully privatized and demonopolized.	• Pursue the reduction in import tariffs further, to achieve a trade-weighted average of 22% in 1998. • Pursue active trade policy to improve market access for Romania food and agriculture products.
C. Taxation	• Agricultural taxes are generally lower than other sectors. Profit tax on primary production is 25%, versus 38% the regular rate. Some agricultural products are either tax except, or benefit from a lower rate of VAT. • An "agricultural revenue tax" was legislated in 1995, but not implemented. The tax is based on the land owned / it is a land tax. However, until 1999 the tax will not be applied. • Informal sector, that accounts of most of the agricultural production, it is not taxed.	• Improve tax administration and tax collection in general. • Increase taxation of informal segments of agriculture while continuing to decrease taxation of formal sector as well as reported personal incomes. • Provide increased tax incentives for investment from properly reported corporate end personal incomes.

ISSUE	STATUS OF REFORMS	OBJECTIVES PROPOSED ACTIONS
2. Land Reform and Farm Restructuring	A land reform was legislated in 1991, and its implementation is close to completion. Amendments to the land reform, to increase the scope of the 1991 restitution, are expected in March 1998. • The land under former collective farms was restituted to the former owners and their heirs, or given to the workers of the collective farms, under the 1991 land law. About 9.3 million ha were or will be restituted to about 4.7 million persons. About 70% of the claimants have definitive titles. • In Oct. 97, the Parliament adopted a Law on the legal circulation of the land that removed the moratorium on the land sales. • About 2.2 million ha, situated mainly in the mountain and hilly areas, was not collectivized during central planning. The owners have full ownership rights over the land, and it is operated as small-scale family farms. • The privately-owned land is organized in small-scale farms and association (formal - commercial company-like, and informal). Out of the total 10.5 million ha 0.8 million are organized into formal associations (averaging 500 ha each), 1.4 million ha into informal associations (about 95 ha each) and 8.3 million ha under peasant households (farms of 2.3 ha each). • The private sector accounts for 80% of the gross agricultural output. • The state-owned farms administer about 1.7 million ha of arable land. At the beginning of 1997, there were about 499 vegetal state farms (with arable land) and 112 animal farms (mainly in pig and poultry production, without arable land). Privatization started in the animal farm sector, but it is blocked in the vegetal farm sector due to unclear ownership of state land. • The Government intends to transfer the use of the state land in private management, using various land tenure arrangements (sale, lease, concessions). Pilot privatization of 50 state farms expected to begin 1998.	Privately owned smaller and larger viable farms are the dominant components of farming system with service and transferable ownership rights. • Accelerate the processing of titling new privatized land. • Amend land ownership and land market regulations by: (a) allow land sales (b) rapidly to allow locators to possess their land; (c) clarify ownership rights of remaining land under state companies so that this may be privatized; (d) allow more flexible land leasing procedures. • Introduce measures to facilitate a speedy consolidation of land ownership and changes in farm sizes. • Develop a strategy for further privatization of remaining state farms.

ISSUE	STATUS OF REFORMS	OBJECTIVES PROPOSED ACTIONS
3. Competitive Agroprocessing and Services for Agriculture.	**Privatization of agroprocessing, input suppliers, storage and services was slow until 1996. An impetus to the privatization process occurred in 1997.** • As of Aug-97, out of a total no. of 534 companies in agro-industries, 184 were privatized. The most advanced subsectors are breweries, milling and baking and edible oil plants. Lagging behind is the privatization of sugar and tobacco industry. • The certified seed business was organized within two commercial companies, producing grain and oilseed varieties, and vegetable seed varieties. The grain certified seed producer is in the process of being reorganized into smaller companies and privatized. The vegetable seed producer now has mixed ownership. • Around 70 storage companies are "successors" of a national parastatal. While 44 of them were included in the mass privatization program and have some dispersed shareholdership, no sale of shares to strategic investors occurred until Oct-97. The Government committed to privatize 50 of them by July 1998. • 70% of the Agricultural Service Companies were privatized, and all of them are advertised for sale. • The upstream and downstream sectors are demonopolized. However, the slow privatization pace in some subsectors combined with the dominance of the MEBO privatization method in others, blocks the infusion of capital in these companies, and maintains obsolete production methods, high costs and significant marketing margins. Competition is at the very beginning in the sectors where the price controls were removed only in Feb-97 (bread, meat, milk and dairy products). • Gross output of agroprocessing (except tobacco) in 1996 is around 65 % of the 1991 level.	**Competitive, privately owned agroprocessing and input supply.** • Accelerate the privatization in animal farms, mechanical service provision, storage, and food industry. • Start the privatization of vegetal farms. • Continue the privatization of food industry. • Implement EU conforming quality and safety standards for agricultural imports and exports. • Improve contract discipline and market transparency. • Promote research and development of new products and markets. • Reform the land reclamation agency (RAIF) and other RAs, turning them into companies that operates on commercial principles.

ISSUE	STATUS OF REFORMS	OBJECTIVES PROPOSED ACTIONS
4. Rural Financing	**An appropriate financial system for privatized agriculture is not in place.** • Up to 1996, most of the agricultural lending was provided through directed credit lines from National Bank of Romania, intermediated mainly through Banca Agricola and backed-up with state guarantees. In 1996, laws and regulations instructed banks to lend to clients that had not repaid previous loans. Despite the preferential interest rates (often with negative real rates), the collection rate for agriculture was worse than in other sectors (60-70%). NBR-supplied credit contributed significantly to inflation. • In 1997, 1050 Bln Lei (150 mil. US$) was placed on the budget and lent for grain planting (550 Bln Lei) and purchase of wheat for domestic bread-making (500 Bln. Lei). The credit was intermediated at marginal real positive rates through 7 banks. Banks were responsible for borrower selection and repayment of the loan at maturity. Bonuses were given to the banks that lent to risky sectors, and to the borrowers if they repay at maturity. Loan collection by Sep-97 was reported to be over 90%. • The Government intends to continue this mechanism in 1998. • Credit cooperatives serve rural household mainly in household credit. There is a draft law for the split-up of the credit from consumer cooperatives. • High interest rates and the lack of collateral seriously limit lending to agriculture. • Restructuring of the Banca Agricola is in process; its bad loan portfolio, estimated at 3500 Bln Lei, is being reincorporated in the public debt.	**Viable financial institutions serving the agricultural sector efficiently.** • Phase out credit subsidies. • Develop banks network serving rural areas. • Promote the emergence of competitive collateral services for agriculture (accept land as collateral; develop the warehouse receipts system; start collateral registration; adopt a good collateral law; develop commodity exchanges) • Develop credit cooperatives and other low-cost financial intermediaries in rural areas.

ISSUE	STATUS OF REFORMS	OBJECTIVES PROPOSED ACTIONS
5. Institutional Framework	**Institutional structure was recently reformed, quality of public services, however, is not in place.** • Ministry of Agriculture renounced to its former roles in price control or supervision and direct distributor of subsidies, for market-oriented functions such policy formulation, extension, research, market information, rural development. • Reorganization of the extension system is planned for the near future. • Information system required by a market based agriculture is only partially in place. • Public investment in agricultural infrastructure is not prioritized on the basis of cost-benefit analysis.	**Promote the use of and support for commercial and private agriculture.** • Complete the reorganization and improve quality of public agricultural administration to the needs of a market economy. • Complete the reform of agricultural extension and research. Establish a sustainable research framework. • Provide assistance in the development of the producer associations, stimulation of farmer marketing and input purchase cooperatives. • Support productive investment in infrastructure for grain marketing and for irrigation.

67

SLOVAK REPUBLIC

Total Population	5.4 mil	Food and agriculture GDP (1995)	5.6%	Agricultural output in 1995 as percentage	
Rural Population	n.a.	Food and agriculture in active labor (1995)	8.5%	of 1987-90 average level	80%
Total Area	4.9 mil ha	Food and agriculture in exports (1995)	6.0%	Livestock production in 1995 as percentage of 1987-1990 average level	65%
Agriculture				Share of livestock in agriculture (1995)	42%
area:	2.5 mil ha	in imports (1995)	8.2%	Agricultural area in private use (1995)	81%
Arable land	65%	Traditionally net importer: grain,		Share of independent private farms in	
Orchards	4%	sugar, vegetable oil and livestock		total arable area (1995)	9%
Forested	39%	products.		Share of private sector in total agricultural output (1996)	86%

ISSUE	STATUS OF REFORMS	OBJECTIVES PROPOSED ACTIONS
1. Macro-economic Framework for Agriculture	**Stable macroeconomic environment and advanced structural reforms. High growth rate of 7.4%, declining unemployment of 10.9% and inflation of 6% (1996 figures).**	**Distortion free market and incentive system.**
A. Prices/Subsidies	• Prices have been liberalized in 1990/91 but some distortions persist through border measures and government subsidies.	• SFMR is turning into an obstacle to market development and ways to make its interventions less frequent and more predictable should be considered.
	• Subsidies declined by 50% since 1990. Measured by the Producer Subsidy Equivalent, aggregate support to agriculture declined from 60% in 1990 to 28% in 1995.	• Gradual expansion of private sector trade activity is needed to improve efficiency. With greater reliance on private trade, modern market-based risk management techniques (e.g., hedging of price risk) would soon develop.
	• State Fund for Market Regulation (SFMR) intervenes to stabilize prices of sensitive commodities such as wheat, pork and poultry.	• Minimum prices should be de-linked from production cost and their level should be gradually reduced.
	• Minimum prices for main agricultural products are set at the beginning of the planting season for specific quantities and quality standards. They cover 90% of estimated average production cost, assuming average yields and taking into account EU prices.	• Dairy quota should be phased out. • Income support payments to marginal areas should be gradually modifies from a per hectare basis to targeted support programs.
	• Milk is regulated by a system of fixed administrative prices combined with a premium to producers and an annual production quota of 900 million liters. Annual surplus of about 25 percent is exported as dry milk with subsidies.	
	• About 30 percent of agricultural subsidies is in direct payments to marginal areas. Other subsidies are scattered in a number of small commodity specific programs.	

ISSUE	STATUS OF REFORMS	OBJECTIVES PROPOSED ACTIONS
B. Trade Policies	• Import protection is limited and gradually phased out following WTO commitments. • Most export and import licenses are automatic and for registration purposes only. • Export licenses are non-automatic for sensitive products such as cereals and flour, sugar, oilseeds and dry milk. • Average weighted import tariff for agricultural products is 6.9% (1992) excluding temporary import surcharge of 10%. Export surcharge phased out on January 1, 1995.	• Export subsidies should be phased out.
C. Taxation	• 6% VAT for agricultural and food products, 23% for other products. • Agriculture benefits from several exceptions to the Slovak tax code and tax reductions (e.g., 50% fuel tax rebate for farm vehicles).	
2. Land Reform and Farm Restructuring	**Transformation of collective farms and the privatization of state farms and most services have been completed. However, it will take more time for the new owners to turn obsolete production facilities into efficient enterprises.** • Following their legal transformation in January 1993, most cooperatives have remained much the same as before. Cooperatives cultivate 70% of farmland. • Land market is dormant because of excessive fragmentation of ownership and high transaction cost. However, land use has not been affected by restitution as most land is leased and remains in large contiguous plots of at least 50 hectares. • Administration of land registration has improved considerably. • Restructuring and ownership consolidation of cooperatives is forced by poor financial performance either through voluntary reorganization or bankruptcy. • Privatization of state farms almost completed.	**Efficient, internationally competitive private farms and an active land market.** • Land ownership in cooperatives is highly fragmented and ways to facilitate consolidation should be considered. One option is to enable active members to swap entitlements for subsidies for payments to buy out land owners. • Ownership consolidation is essential for reducing land transaction cost, activation of the land market, and acceptance of land as collateral.

ISSUE	STATUS OF REFORMS	OBJECTIVES PROPOSED ACTIONS
3. Competitive Agroprocessing and Services for Agriculture.	Privatization of agroprocessing and services has been completed. New owners will have to reduce cost and improve efficiency to maintain their market share against growing foreign competition. • Most agro-industries and services struggle with weak management, growing debt, and declining labor productivity. • Organized wholesale and retail markets for agricultural commodities and food are yet to develop. Food distribution systems are inefficient, transaction cost are high. For example, gross margins for basic commodities such as wheat are at least twice as high as in developed market economies. • Foreign participation in agroprocessing privatization has been modest.	Competitive and privately owned agroprocessing and input supply. • Continue to resist pressure by special interest for protection against foreign competition. • Continue to improve the enabling environment for business activity.
4. Rural Finance	High risk stemming from undercapitalization, ongoing consolidation of ownership, high debt, and poor credit history, all mitigate against better access and more favorable credit terms in agriculture. • About 80% of financial needs in agriculture is covered from own-resources and 20% by commercial credit. • State Support Fund for Agriculture and Agro-industries (SSFAA) supports medium- and long-term investment in agriculture and agro-industries by interest rate subsidies (5% annual rate) and credit guarantees. • Working capital financing is offered against pledges on future crops guaranteed by commercial banks through forward supply contracts with processors. • SFMR is a source of working capital financing as farmers can sign contracts for future delivery against an advance to finance inputs. • Preparation of legal amendments and new laws is underway to improve collateral law including land mortgage law.	Access to credit should be improved using market-based instruments and techniques, and an efficient universal banking system. • Use of land as collateral is vital for agriculture to obtain better access to long-term investment financed by bank credit. • Speed up bankruptcy procedures and improve protection of creditors in cases of credit default. • Phase out interest rate subsidies. • Promote the emergence of competitive insurance services for agriculture.

70

ISSUE	STATUS OF REFORMS	PROPOSED ACTIONS OBJECTIVES
5. Institutional Framework	**Consolidation and adjustment of main agricultural institutions has been largely completed.** • Number of staff in agricultural and food industry research establishment declined by 50% between 1990 and 1995. • Excessive but inefficient research facilities have been either closed or transformed into consulting services on a commercial basis. • Academy of Agricultural Sciences has taken over a coordinating and financing role in research. • A new concept for agricultural education is under preparation by a joint effort of the Ministry of Agriculture and the Chamber of Agriculture. • Institutions to monitor and enforce quality and health standards have been made more efficient but more needs to be done to meet strict EU requirements.	**Best practices in other countries should be applied, where appropriate, to improve efficiency in the provision of "public goods" to agriculture.** • No specific recommendations have been made.

SLOVENIA

Total Population	2.1 mil	Agriculture in GDP 1995	4.4%	Agricultural output in 1995 as percentage of 1990 level	110%
Rural Population	48%	Food and agriculture in active labor (1995)	10.4%	Livestock production in 1995 as percentage of 1990 level	90%
Total Area	2 mil ha.	Food and agriculture in export (1995)	4%	Share of livestock in agriculture (1995)	47%
Agriculture area:	0.9 mil ha.	Food and Agriculture in imports (1995)		Arable area in private use (1995)	100%
Arable land	28%		8.5%		
Orchards	0.1%	Traditionally net exporter of hops, wine, fruits, beef, eggs, and poultry.		Share of independent full and part-time family farms in total arable area (1995)	90%
Forested	30%				

ISSUE	STATUS OF REFORMS	OBJECTIVES PROPOSED ACTIONS

1. Macro-economic Framework for Agriculture
 	
A. Prices/Subsidies
	The macroeconomic and trade environment in Slovenia is widely liberalized. Import protection and state intervention still cause major distortions for agriculture and food industry.
	• Parts of price policy heritage from former political and economic system
	• Internal markets for grain (wheat and rye), sugar, and milk not liberalised
	• Total budgetary support amounted to 15% of net value added of agriculture in 1994
	• Overall PSE coefficient sums up to 40%
		Redesign trade policy for agro-food products according to WTO requirements. Abolish State monopolies. Prepare Agricultural Policy for EU Accession
		• Improve market structure and foster the development of market functioning
		• Speed up full liberalisation of domestic agricultural and food markets
		• Support product marketing and support establishing regional and national trade marks
		• Focus on direct support instead of price support
		• Prepare CAP implementation according to CAP 2000 Reform:
		  – set up a lean market intervention agency,
		  – stepwise abolish state grain and sugar monopoly,
		  – liberalise milk retail market,
		  – organise price and market regulation in a transparent and systematic way

B. Direct Support and Structural Policy *Compensation Payments* *Investment Policy*	• A systematic approach is lacking • Product specific subsidies do not exist • Input subsidies for different agricultural products are paid • Investment promotion is based on interest rate subsidies, which are not well targeted • Selection of areas and criteria for support are sometimes contradictory	• Shift from price intervention to direct support by 3–5 years budget programs for agriculture (like US farm bill) • Abolish stepwise input subsidies • Introduce lump sum payments per ha in combination with maximum livestock per ha (2 – 2.5 LU), based on defined programmes, • Assess the "need" of compensation payments for farmers during CEFTA liberalisation and EU Accession • Clearly define objective of investment programs • Limit State support on investments improving agricultural competitiveness

ISSUE	STATUS OF REFORM	OBJECTIVES PROPOSED ACTIONS
C. Trade Policies	• Border Protection is still the most important instrument, with high tariffs and some non-tariff barriers. • Main Trading partners are former Yugoslavia and EU • Since October 1994, Slovenia is member of GATT and became a founding-member of WTO • A free trade agreement has been signed with EFTA countries	• Actions to be carried out simultaneously: • Fully comply with WTO commitments, • Preparation for CEFTA agricultural Free Trade in 2000, • Preparation for EU accession
2. Land Reform and Farm Restructuring	**Land privatized but land market not fully developed** • Average private farm size is four hectare • Dispersed structure of plots • Land prices approach the highest levels in Europe • Cadastral register is incomplete hindering land mobility • Limited mobility of land	**Increase land mobility to improve farm sizes** • Use land taxation to encourage land consolidation • Restructure state land agency • Complete land register as soon as possible
3. Competitive Agroprocessing and Services for Agriculture *Farm Profitability* *Agricultural Co-operatives* *Agro-industries*	**Enterprises are partly efficient and internationally competitive, but vestiges of old system hinder progress** • Farm income is mostly much below the defined "parity income level" • Widespread over-mechanisation • Lacking knowledge of modern marketing and business practices • Lack of co-operation in product marketing and input purchase • Weak economic performance of co-operatives with limited value added • Status of co-operatives are one major constraint to develop competitive up- and downstream sectors • Operating in a monopoly or oligopoly structure	**Improved farm efficiency; transformation of co-ops and preparation for EU accession.** • Promote joint use of machinery (machinery circle) • Start information campaign on cost reduction and productivity improvements in the different production areas • Support production of high quality instead of mass production • Revise the legal and organisational framework for co-operatives • Transform existing co-operatives to real marketing co-operatives • Support specialisation and market activities of agricultural co-operatives • Provide information on legal, hygienic, and health standards of EU • Support vertical and horizontal integration • Increase product quality • Accelerate move to EU quality standards

ISSUE	STATUS OF REFORM	OBJECTIVES PROPOSED ACTIONS
4. Rural Financing	General banking system is active in rural lending but the general problem of high interest rates is affecting rural economy • Agricultural Credit co-operatives focus too narrowly on agriculture. Mostly active for disbursement of state support • Weakness: they do not add significant value added in terms of financial intermediation to the agricultural and rural sector • Lacking investment possibilities in agriculture is less a phenomena caused by missing credit resources than economic situation of farming and small structure of production	Fully functioning and viable rural financial system • Analyse the real bottlenecks of the rural financial system • Assist farmers and other rural entrepreneurs in defining viable investment projects • Support the change of approach from an asset based to a concept (business plan) based lending approach
5. Institutional Framework	Institutional Framework is slowly adjusting to full market-based conditions. • Internal and external communication and Co-ordination is weak in governmental work • Centralisation of public activities (which is the main objective of administration) is the wrong approach • Responsibilities and division of tasks between MAFF and MOERD are unclear • The representation of professional interest in the policy making process is weak • Independent NGOs are not covering the full spectrum of agriculture and food activities • Role of State and public services in agriculture is not clearly defined	Efficient and effective public sector administration and support for commercial and private agriculture. • Clearly define responsibilities for agricultural policy and the role of the MAFF and MOERD • Strengthen the professional and administrative capacities of the MAFF • Outsource, there appropriate, selected services from MAFF • Contracting specific tasks (information, food quality, policy analysis and preparation of EU negotiations) • Strengthening of MAFF's staff and NGOs to address the dialogue with foreign partners in Slovenia • Develop efficient procedures for the dialogue with EU and other international partners in order to reduce paper work and communication costs • Involve professional organisations, universities, NGOs in consultation, organisation and implementation of public activities in the field of agriculture and rural development • Take care that public administration is in balance with the size of the sector • Clearly define the role of the planned agricultural chamber • Establish a multi- institutional but independent agricultural policy analysis and rural development centre, with high degree of flexibility

Commonwealth of Independent States

Policy Matrices

ARMENIA

<u>Total</u> <u>Population</u>	3.7 mil.	Food and agriculture in NMP* 1996	37%	Agricultural output in 1996 as percentage of 1988 level 86%
<u>Rural</u> <u>Population</u>	31%	Food and agriculture in active labor (1996)	37%	Livestock production in 1996 as percentage of 1988 level 64%
	3 mil ha	Food and agriculture in		Share of livestock in agriculture (1996) 38%
<u>Total Area</u>	1.4 mil	exports (1996)	5%	Agricultural area in private use (1996) 32%
Agriculture area:	ha	in imports (1996)	37%	Share of independent private farms in
		Traditionally net importer:		total arable area (1996) 65%
	35%	grain, sugar, vegetable oil		Share of private sector in total
Arable land	5.5%	and livestock products.		agricultural output (1996) 99%
Orchards	15%			
Irrigated	11%			
Forested				

ISSUE	STATUS OF REFORMS	OBJECTIVES PROPOSED ACTIONS
1. Macro-economic <u>Framework for</u> <u>Agriculture</u> A. Prices/Subsidies	**Market liberalization is advanced, although some delays in full completion.** • Agricultural producer and consumer prices were deregulated in 1992. • State orders (state tasks) set for grain and some other major products were eliminated in 1995. • Subsidization of agriculture has largely been discontinued, however subsidization of irrigation water still exists. • Domestic grain price adjusted to border price. • Underdeveloped markets and limited demand keep producer prices, except grain, under border prices. • Profit and marketing margin control in the food processing industry has been removed.	**Distortion free marketing and incentive system.** • Phase-out all subsidies to irrigation, and increase water charge recovery rates up to 80-90%. • Discontinue preferential sales of grain and flour to state-owned grain milling and baking enterprises. • Introduce an adequate "social safety net" of subsidies targeted to low income and vulnerable consumers.
B. Trade Policies	• Grain and other commodities for state reserves are procured by commercial methods. • Export ban on grain products removed. • Food exports no longer require licenses. • Tariffs reduced to low and uniform rates.	• Proceed with demonopolization, privatization of remaining state trading enterprises. • Pursue active trade policy to improve market access for Armenian food and agricultural products, especially in republics of the FSU and Central and Eastern Europe.
C. Taxation	• 20% VAT (farms are exempt). • Over-taxation of food processing by a rather complicated system.	• Larger private farms should be incorporated into the regular business tax system. • Simplify and reduce taxation in agroprocessing and promote investments in tax incentives.

---

* Net Material Product

ISSUE	STATUS OF REFORM	OBJECTIVES PROPOSED ACTIONS
2. Land Reform and Farm Restructuring	**The most comprehensive land reform in FSU transferring most of arable and perennial crop areas to private farmers.** • Agriculture was decollectivized in 1991. • About 20% of arable land is kept in state reserve. • Reserve land is utilized by leasing and is to be sold to private farmers. • Pastures and meadows remained largely state and municipality owned, with some being sold while most are leased. • Land sales are allowed; however, market for land and leasing is developing slowly. • Law creating legal conditions for a land market was adopted in December 1995. • Technical conditions (titles, registration system) for a functional land market are not in place. About 100,000 titles were issued by early 1998.	**Individual private farming is the predominant structure in the farming system, with secure and transferrable ownership rights.** • Develop and implement a property registration, and information system to provide security of tenure, full information on property transactions, and a basis for real estate taxation. • Lengthen lease periods and privatize reserve land. • Prepare and implement a program to promote the emergence of land markets to support land consolidation and the move towards a more efficient holding structure. • Rearrange responsibilities for providing social services in rural areas and guarantee the continuation of rural social services after transfer. • Establish competitive land mortgage and credit systems. • Create a framework conducive to organizing local and regional service cooperatives.
3. Competitive Agroprocessing and Services for Agriculture.	**Delays in privatizing agroprocessing and services for agriculture.** • Agroprocessing privatization was not part of the original agricultural reform package of 1991. • The privatization of agroprocessing and input supply is part of overall privatization program that was accelerated significantly in 1995-1996. • Out of 220 agroprocessing plants, 200 have been privatized. • The grain industry is covered by current privatization programs. Most bakeries were privatized by 1997. • Foreign participation in the agroprocessing privatization is modest.	**Competitive, privately owned agroprocessing and input supply.** • Complete privatization of agricultural processing. • Establish feasible and reasonable quality and safety standards for agricultural imports and exports. • Acquire technical assistance and training in enterprise management. • Promote joint ventures to tap foreign expertise, technology, capital, and provide access to foreign markets. • Promote research and development of new products and markets.

ISSUE	STATUS OF REFORM	OBJECTIVES PROPOSED ACTIONS
4. Rural Financing	**Lack of an appropriate financial system for privatized agriculture.** • Financing in agriculture is not adjusted to the needs of a market based privatized agriculture. • High interest rates and the lack of a registration system restricts the use of land as collateral and seriously limits lending to agriculture. • Restructuring of Armgrobank is in process. • Establishment of the Agricultural Cooperative Bank of Armenia (ACBA) is complete and rural lending has been initiated through dollar denominated loans and in local currency.	**Viable financial institutions serving the agricultural sector efficiently.** • Do not use fiscal means involving financial institutions to sustain the operation of state enterprises critical to food security. • Promote growth of ACBA and other rural savings and credit societies. • Promote the emergence of competitive insurance services for agriculture.
5. Institutional Framework	**Adjustment of institutional structure is slow and constrained by budgetary difficulties.** • New Ministry of Agriculture and Food Industry was created. • Education system has been partially adjusted to emerging new conditions. • Reorganization of the research system is currently taking place. • Public activities (government research-education) in agriculture are seriously hampered by budgetary difficulties. • Armenian agricultural extension system was created but does not cover the whole country yet.	**Efficient and effective public sector administration and support for commercial and private agriculture.** • Complete the reorganization of public agricultural administration to the needs of a market economy. • Complete the reform of agricultural education and research. • Promote the establishment of a system to provide for technical assistance for enterprise restructuring.

AZERBAIJAN

Total Population	7.5 mil				
Rural Population	44%	Food and agriculture in GDP 1994	27%	Food production index in 1995	63
Total Area	8.6 mil ha	Food and agriculture in active labor force (1994)	38%	(with 1989-91 levels=100)	
Agriculture area:	4.4 mil ha				
Arable land and		Traditionally net exporter of			
Permanent crops, incl.	45%	cotton, fruits and vegetables to former Soviet Union Countries			
Irrigatible land	20%				
Permanent pasture	50%				
Forested	13%				

ISSUE	STATUS OF REFORMS	OBJECTIVES PROPOSED ACTIONS
1. Macro-economic Framework for Agriculture A. Prices/Subsidies	After several years of delay a rapid process of agricultural reforms has been followed since 1996. Key reforms to date include: • complete abolition of the state order system • the break-up of most state and collective farms and distribution of arable land to households • distribution of most livestock to households • removal of quantitative controls on external trade in agricultural products • domestic price liberalization and the privatization of most agro-industrial enterprises through direct, voucher and auction sales	A distortion-free, efficient and internationally competitive agricultural sector • Agricultural sector policies within macroeconomic policy framework aimed at limiting the upward movement of the exchange rate (Dutch disease) to maintain agricultural sector competitiveness • Define extent and mechanisms for subsidization of the irrigation system within context of establishment of a cost-recovery system
B. Trade Policies	• The foreign trade regime and the associated payments systems has been largely liberalized, leading to the development of world-market related domestic price structures for most commodities • cotton exports for the 1997 crop still channeled through Agroincom • strong pressure for protection of the wheat producing and processing sectors as imports depress domestic prices.	• Adopt a consistent trade regime based on international competition with tariffs limited to anti-dumping (genuine cases only) and a low uniform revenue generating tariff to be eliminated as oil revenues supplement the government budget • immediate independent assessment of current wheat and wheat flour imports to examine dumping claims • appoint independent external inspectors to verify external trade quantities and prices • pursue longer term aim of joining WTO

ISSUE	STATUS OF REFORMS	OBJECTIVES PROPOSED ACTIONS
C. Taxation	• Inputs and outputs subject to 20% VAT • Farmers with incomes above a minimum threshhold pay income tax under a progressive rate structure. The top marginal rate is 40%. • Land tax is based on the region, quality of land, and use (arable, fallow, perennial crops).	• Government to develop a medium and longer term fiscal strategy for the agricultural sector defining the planned change in the net resource flow to the sector as oil revenues come on stream.
2. Land Reform and Farm Restructuring	**Government committed to transforming agriculture into an efficient and dynamic sector by encouraging the development of a market-based, predominantly privately owned production system, with rapid progress in land distribution in 1998.** • Significant progress in the privatization of agricultural land. State Land Committee and Land Institute have made substantial progress in establishing a Lands Registry which issues legal land titles and constitutes the national land cadastre. • Effective Distribution began after Land Reform Law was passed (July 1996) with accelerated speed from January 1997 onwards. Data processing demands of the system are extensive and are creating delays in the issuing of titles. • Apart from a 5% "land reserve" all arable land is to be distributed to private farmers. All other lands including pastures are allocated to the state and the municipalities. An efficient system of future pasture management is yet to be established. • By end-October 1997, 28,000 farms (with an average of 4 family members) representing 80,000 ha had gained legal title. A much larger number is already farming their assigned plot of land. • Resulting farm sizes are very small and often not viable to support farm families. • The system of land quality valuation and distribution is not always considered fair. The same applies to the distribution of assets where distribution procedures sometimes favor rent-seeking behavior of ex farm managers. Even after land and asset distribution the incentives are strong to remain in some transitional form of collective farm. • There is no evidence of a functioning lease market for land.	**A working land market for efficient and fair asset transfer and a socially acceptable land consolidation process.** • Further acceleration of land registration and titling services as a priority to developing a fully functional land market. • Establish a framework for consistent land and asset valuation and for the dissemination of land market information. • Strengthen the legal and institutional basis for the leasing of land. • Improve the design of the legal and institutional framework that provides for financially sound mortgage operations. • Establish efficient arrangements for the sustainable management of state and municipal lands i.e. pastures and forests. • Ensure equal rights for farmers within and outside of transitional collective farms and abolish disincentives to farm independently. • Promote new roles for western type cooperatives in agriculture.

ISSUE	STATUS OF REFORMS	OBJECTIVES PROPOSED ACTIONS
3. Competitive Agroprocessing and Services for Agriculture	**Privatization of agro-industry is proceeding rapidly, with major industries such as the State Bread Concern and the cotton processing sector almost fully privatized.** • Slower privatization progress in some remaining sub-sectors, including tobacco, fruit canning, wine and spirits • Permissive approach to foreign investment, but actual investment mainly in cotton ginning to date • Little progress yet in recovering market share in FSU markets lost since 1992 • Major industries, including grain milling and processing have over-capacity, especially in light of high growth of imports of grain and flour products, leading to strong political pressures for protection • Basic legal framework for private sector agro-industries still highly inadequate, inhibiting access to bank credit • Inadequate or non-existent market information and support services for producers and the marketing chain	**Reestablish an export led diversified product base in agriculture. This will require rapid productivity growth in the agro-processing sector to regain markets and to offset the expected upward movement of the manat as oil revenues increase.** • Complete privatization agro-industry to include productive capacities currently retained in public sector (cattle breeding, seed multiplication) • Adopt a permissive attitude to post-privatization rationalization in key over-sized industries such as grain processing • Further develop the framework for attracting foreign investment in agro-processing • Sell remaining state shares in agro-processing • Create a support service to accelerate private sector investment in agro-processing • Review cotton industry after first full year of privatization to determine if there is a need for regulatory framework to control a cotton processors oligopsony • Azerbaijan State Wheat Reserve Agency given private sector supply capacity. If possible abolish the Agency.

ISSUE	STATUS OF REFORMS	OBJECTIVES PROPOSED ACTIONS
4. Rural Financing	The former rural credit system has collapsed, with the main institution, Agro-Prom Bank, under Central Bank led restructuring. There are no local providers of longer term credit for the sector • Emerging privatized farms have neither the experience nor the asset base for borrowing from financial institutions • Some signs of processors providing crop input credit, especially in the cotton sector • The commercial banking sector is itself going through a rapid process of restructuring with the number of operating banks reducing sharply. Few banks lend to the agricultural sector, preferring the higher and less risky returns in trade and oil industry financing • No existing basis of grass-roots savings and credit institutions and recent attempts to start credit unions have met legal obstacles	A private sector-based rural financing system, based on indirect channels of credit to producers via processors and input and service suppliers rather than on direct credit channels to household farms. • Continue the restructuring program for AgroProm Bank and clarify its post-restructuring role in rural finance. Define its policy for future lending to the farm sector (if any) • Central Bank to establish clear legal framework for credit unions and other grass-roots savings/credit organizations • Improved supervision and regulation of the commercial banking sector and measures to ensure compliance by participating commercial banks in on-lending schemes for the agricultural sector • Government to avoid a top-down approach to directed development of the credit union sector and to allow gradual development based on active grass-roots participation
5. Institutional Framework	Re-organization of MOA under preparation; however, other state organs with a role in agriculture (irrigation and veterinary state committees in particular) not reformed yet and weakly coordinated with MOA. • New institutions needed to manage irrigation and drainage networks. • Sector still lacks a capacity for policy making and for producing reliable statistics on private agriculture. • Most production and commercial functions in agriculture have been removed from government. However, this has not occurred in seed multiplication, livestock breeding and veterinary services. • Farmers face insufficient support services, particularly regarding technical & management advice as well as market information. • Institutions responsible for monitoring/regulating natural resource management, plant/animal diseases and trade/use of related drugs have not adjusted to the context of private farming. However, new laws have been passed on seeds, plant/animal property rights, and plant protection.	Review the role of government, notably in key subsectors such as seed production, livestock development and irrigation. For seed and livestock, this could be part of the preparation of development strategies. • A subsector review is long overdue for irrigation. • Establish a strong capacity in MOA or Cabinet of Ministers for policy making; develop new statistical instruments. • Develop an essentially private farm advisory service and market information system, first on a pilot basis at regional level. • Develop a strategy for the development of farmer-oriented adaptive research and the restructuring of the national agriculture research system. • Review the legal/regulatory framework on pest/pesticide control, property rights, veterinary medicine, etc. Review existing strategies for controlling quarantine pests & diseases.

BELARUS

Total Population	10.3 mil.	Agriculture and Forestry in GDP (1995)	24%	Agricultural and forestry output in 1994 as percentage of 1990 level — 72%
Rural Population	32%			
		Agriculture and Forestry in active labor (1996)	20%	Livestock production in 1995 as percentage of 1990 level — 57%
Total Area	21 mil ha.			
Agriculture area:	9.4 mil ha	Food and agriculture in exports (1994)	25%	Share of livestock in agriculture 1994 — 42%
				Agricultural area in private use (1996) — 16%
Arable land	67%	in imports (1994)	5%	Share of independent private farms in total agricultural land (1996) — 1%
Orchards	2%	Traditionally net exporter of livestock products, potatoes, and flax.		
Pastures/meadow	33%			Share of private sector in total agricultural output (1996) — 40%
Forested	36%			

ISSUE	STATUS OF REFORMS	OBJECTIVES PROPOSED ACTIONS
1. Macro-economic Framework for Agriculture A. Prices/Subsidies	**Agricultural markets are Government controlled to considerable extent.** • Progress in reforming the agribusiness sector has been made at the national level by reducing and removing administrative controls over pricing, processing, procurement and distribution. • State procurement of major agricultural products at cost based recommended prices is still in place. • Producer prices have been decontrolled; however, government intervention in selected agricultural markets remains strong. • Accomplishments at the national level are still incomplete and are being undermined at the regional or Oblast level by the local administration. • Some remaining floor and indicative prices are often enforced by regional authorities as mandatory prices. • Significant subsidization of agriculture (about 7% of GDP in 1996).	**Competitive and functioning agricultural markets without Government intervention.** • Place any necessary state procurement on the base of competitive open bidding. • Continue with liberalization of all prices and eliminate regional interference from agricultural markets. • Measures are needed to loosen the hold of the regional administrations and their agriculture departments over enterprises in the sector, in order to ensure that the state's laws on price liberalization are respected. • Phase-out producer subsidies. • Eliminate all formal and informal barriers in domestic trade. • Ensure food supply and a minimum safety net for the poor.
B. Trade Policies	• Export quotas with licensing for many plant products, especially grains and rapeseed. • Minimum export price for selected food products were introduced beginning 1995. • Customs union with Russia abolished all border controls and increased import duties to non-CIS countries by 15% points on average. • VAT on exports have been eliminated for exports to non-CIS countries. • Surrender requirement exists for 50% of export earnings.	• Refrain from intervening in agricultural import and export markets with the exception of interventions acceptable under the WTO. • Eliminate minimum export price requirement. • Introduce low and uniform tariffs to all countries and remove quantitative restrictions on exports. • Break up monopolistic trading organization • Privatize all trading enterprises to eliminate implicit de facto price controls by state owners.

ISSUE	STATUS OF REFORMS	OBJECTIVES PROPOSED ACTIONS
C. Taxation	• Overall tax on wages of approximately 40% is almost the same as in other sectors. • Profit tax rate of up to 30% for non-agricultural businesses like food industries. Half reduction of rates for profits of small enterprises with less than 25 employees or less than 10 in trade sector and for profits from exports (all enterprises). • 15% payment from profits to the Centralization Fund of the Ministry of Agriculture. • Land tax of about US$ 0.6 - 3.5/ha is adjusted annually.	• Guarantee the consistency of agricultural profit tax with the rest of businesses like food industry. • Fair and non-discriminatory system of taxation.
2. Land Reform and Farm Restructuring	**Land reform and farm restructuring are still at a very early stage.** • Restructuring and privatization of large-scale farms moves very slowly (but these farms produce major share of output of selected products). • Most of state farms were transformed into collective agricultural enterprises without changing the mode of operation. • Number of small private farms is minimal and increasing slowly (less than 1% of cultivated land belongs to them). • Housing and household land, about 15% of total agricultural land, was given for full private ownership. • Private land ownership is maximized at 1 ha per person, however, can freely be traded. • Rest of agricultural land remains in state ownership. • Government intends to provide long-term lease rights for the large farms and not for the individual members of these farms, as well as for individual farms. • About one third of large farms are in serious financial condition. • Land titling and registration practice required by a functioning land market does not exist, i.e. there are major impediments to the transfer of land.	**A farming system based mainly on private ownership of land and a working land market for efficient and fair asset transfer.** • Develop a consistent framework for restructuring and privatization of large collective and remaining state owned farms. • Develop and implement a consistent policy to privatize the land of large farms. • Support emerging private farming and guarantee fair conditions for those who wish to start individual farming. • Allow downsizing of livestock sector. • Create the legal and technical conditions for a functioning market for agricultural land. • Develop mortgage procedures for land, other real estate, and moveable assets. The mortgage law would allow lessees to mortgage their leasehold interest. • Adopt a resolution to establish a single registry of land and other real estate. • Develop an enabling environment to stimulate increased off-farm employment in rural areas. • Safeguard an acceptable level of rural social services during the period of restructuring and privatization of large-scale farms.

ISSUE	STATUS OF REFORMS	OBJECTIVES PROPOSED ACTIONS
3. Competitive Agroprocessing and Services for Agriculture	**Little progress, as major privatization program has yet to come.** • Nearly all of the enterprises in the agriculture input and food marketing subsectors are still owned and operated by the state-owned companies or by the local municipality. • The privatization of agroprocessing and input supply industries is progressing slowly. • Many enterprises have been corporatized and a few also privatized. • The stated objective was to complete the corporatization process in agro-industry by 1996. • Under the existing privatization scheme, collective and state farms shall become major shareholders of food processing plants. • State procurement at unprofitable rates still in place for selected commodities. • Minimal foreign participation in the process of privatization.	**Competitive, privately owned processing, input supply and service subsectors.** • Implement an overall program of privatization without delay in agroprocessing and input supply. • Demonopolize state corporations (associations) and privatize them individually by plant or unit. • Open participation in privatization for all investors. • Adopt anti-monopoly legislation. • Facilitate foreign investment by implementing foreign investment laws. • Remove all explicit and implicit price controls in concert while assuring a competitive market structure.
4. Rural Financing	**Existing financial system subsidizes the agricultural sector.** • Both primary agriculture and agroprocessing have serious liquidity crises and rely on Government credit and directed credit from the banking system. • Effective private agricultural banking system is absent.	**Viable financial institutions efficiently serving the agricultural sector.** • Prepare an action plan to revitalize financial services. • Implement pilot projects to establish village credit unions. • Restructuring of Agrobank is needed. • Phase-out subsidized credits to agricultural producers.

ISSUE	STATUS OF REFORMS	OBJECTIVES PROPOSED ACTIONS
5. Institutional Framework	The role of the government in agriculture has not changed appreciably since the initiation of the reform. • Only minor changes in the Soviet type Government structure related to agriculture. • Research/education system has not been adjusted to emerging new conditions. • Public activities (government research-education) in agriculture are seriously hampered by budgetary difficulties. • Western type agricultural extension system does not exist.	Efficient and effective public sector administration and support services. • Prepare and implement a program of re-organization of public administration in agriculture. • Simplify the structure of governmental organizations corresponding to the reduced role of the public sector responsibilities in agriculture. • Review the agricultural education and research system. • Support the emergence of private farm advisory services. • Strengthen infrastructure and transportation systems in the rural area.

GEORGIA

Total Population	5.4 mil	Food and agriculture in GDP (1997)	28%	Agricultural output in 1997 as percentage	
Rural Population	44%	Food and agriculture in active labor (1993)	33%	1989 levels	43%
				Livestock production in 1997 as	
Total Area	7 mil ha.	Food and agriculture in exports		percentage of 1987 level	64%
Agriculture area:	3 mil ha.	(1996)	16%	Share of livestock in agriculture (1994)	44%
		in imports (1996)	25%	Agricultural area in private use (1996)	24%
Arable land	26%	Traditionally net exporter of wine,		Share of private sector in total	
Orchards	11%	processed and fresh fruits,		agricultural output (1994)	85%
Irrigated	45%	vegetables and tea			
Forested	40%				

ISSUE	STATUS OF REFORMS	OBJECTIVES PROPOSED ACTIONS
1. Macro-economic Framework for Agriculture A. Prices/Subsidies	**Liberal agricultural markets mainly free of Government intervention.** • Producer prices have been liberalized and minimal government intervention on agricultural markets. • Control of energy prices is being gradually readjusted to reach cost recovery levels. • Subsidization of water charges is the last remaining producer subsidy.	**Competitive and functioning agriculture markets, without Government intervention.** • Abstain from direct price interventions. • Phase out remaining producer subsidies (water).
B. Trade Policies	• State order system for agricultural products was abolished in late 1995. • 12% uniform tariff on imports. (Trade with CIS countries is duty-free). 5% on selected capital goods, raw materials, and medicines. • Tax on exports was eliminated in late 1994. • Liberal export policy with no licensing for most agricultural products. • Most agricultural products are traded on essentially private, informal markets. • Illegal rent seeking and bureaucracy seriously constrains domestic markets.	• Refrain from intervening in agricultural import and export markets with the exception of interventions acceptable under the WTO. • Proceed in the process of accession to the WTO. • Ensure that all state agricultural procurement is executed on a competitive basis. • Maintain low and uniform tariffs and no quantitative restrictions or taxes on exports. • Improve physical and commercial infrastructure for export trade rather than introducing special incentive programs. • Introduce provisions for bonded warehouses and duty drawbacks to promote exports.
C. Taxation	• Credit for VAT payment for capital goods established in June 1995.	• Guarantee the consistency of land tax with the rest of taxation. • Simplify taxation of agroprocessing and services. • Fair and non-discriminatory system of taxation.

ISSUE	STATUS OF REFORMS	OBJECTIVES PROPOSED ACTIONS
2. Land Reform and Farm Restructuring	**Progressive, but rather spontaneous, unstructured and unfinished land reform.** • The 1992 land privatization program provided 61% of arable land and 81% land for perennial crops but remained unfinished. • Law giving ownership rights to the beneficiaries of the 1992 land reform was passed by the parliament in early 1996. • Remaining large scale farms still control about 48% of arable and perennial crop areas though large part of this land is already leased out to private farmers. • The current state of large scale farms is beyond the point when they can be usefully restructured. • Land titling and registration practices required for the functioning of the land market do not exist, but legal framework is in place. • Law on providing lease rights to state owned agricultural land was enacted in mid 1996.	**Private farming as the major component of the farming system with secure transferable land use rights.** • Complete the initial privatization of land as envisaged by the Presidential decree of December 1992. • Create the technical conditions for a functioning land market through the establishment of title registries. • Transfer remaining cultivated state lands to private producers. • Transfer non-land productive assets from state/collective users to private users and operators. • Develop mortgage procedures for land, other real estate, and moveable assets. • Adopt a resolution to establish a single registry of land and other real estate.
3. Competitive Agroprocessing and Services for Agriculture	**Spontaneous and slow privatization program.** • The early phase of privatization was rather spontaneous and often inequitable. • Currently agroprocessing and input supply industries are privatized in the framework of the overall privatization program. • Out of 457 state owned agro-industrial enterprises, 440 had been corporatized and began to privatize by December 1997. • Fifty-one agro-industrial enterprises remain in state ownership. • Most of agro-industries and also those privatized are operating at low capacity at best, due to energy shortage and the lack of liquidity. • Bakeries and mills have been privatized. • Foreign participation in the process of privatization remains low, but there are growing signs of interest. • Several service enterprises were privatized as of October 1995 but have not changed their mode of operation. No competing organizations have been created in the process of privatization.	**Competitive, privately owned processing, input supply and service subsectors.** • Implement the overall program of privatization without delay in agroprocessing and input supply. • Create transparent ownership of privatized agroprocessing firms as quickly as possible. • Demonopolize state corporations (associations) and privatize them individually by plant or unit. • Reduce the number of agroprocessing enterprises remaining in state ownership to the minimum, even in the short term. • Facilitate the emergence of new and restructured private firms in processing, input supply and services. • Encourage improved product quality and penetration of domestic and foreign markets.

ISSUE	STATUS OF REFORMS	OBJECTIVES PROPOSED ACTIONS
4. Rural Financing	Existing financial system is at a rudimentary state of development and does not serve the agricultural sector. • Both primary agriculture and agroprocessing have a serious liquidity crisis. • A major coordinated effort is underway, supported by the World Bank, IMF and bilateral donors to strengthen the banking system's infrastructure. • Major restructuring of Agrobank is needed. • About 100 Rural Credit Unions have been formed.	Viable financial institutions efficiently serving the agricultural sector. • Complete bank certification program and maintain restrictions on uncertified banks. • Develop a strategy for the financing of agriculture relying on links between the credit and product. • Continue establishment of credit unions.
5. Institutional Framework	Institutional structure needed by privatized agriculture is not in place. • At the regional level the administrative structure of the socialist era has been preserved with little change. • Research/education system has not been adjusted to emerging new conditions. • Public activities (government research-education) in agriculture are seriously hampered by budgetary difficulties. • Western type agricultural extension system does not exist, but pilot projects have been implemented and are expected to expand in the future.	Efficient and effective public sector administration and support services. • Prepare and implement a program to alter the structure and scope of government organization for management of agriculture. • Re-orient Government attention toward private agriculture. • Re-orient Government role from direct intervention to establishing the general rules and facilitating conditions for the smooth operation of the markets and independent business organizations. • Support the emergence of private farm advisory services. • Restructure the agricultural education and research system.

KAZAKHSTAN

Total Population	16.7 mil	Food and agriculture in GDP 1996	12.3%	Agricultural output in 1996 as percentage of 1992 levels:	53%
Rural Population	29 mil	Food and agriculture in active labor force (1996)	21.8%	Crop production	90%
		Food and agriculture in export (1996)	11.8%	Livestock production	28%
Total Area	170 mil ha	in import (1996)	13.2%	Share of livestock in agriculture (1995)	39%
Agriculture area:	216 mil ha	Traditionally net exporter of wheat and livestock products		Agricultural area in private use (1996)	75%
Arable land	16%			Share of family private farms in total agricultural output (1996)	20%
Orchards	0.5%				
Pasture	84%				
Forested	3%				

ISSUE	STATUS OF REFORMS	OBJECTIVES PROPOSED ACTIONS
1. Macro-economic Framework for Agriculture	**Significant but incomplete liberalization of agricultural markets**	**Competitive markets for outputs and inputs of agriculture with minimal Government intervention,**
A. Prices and Subsidies	• Government price controls ended in 1994/95; • Domestic prices largely follow world market prices though they still reflect some distortions due to underdeveloped local markets; • Purchase price set for Government procurement is still often used as reference price; • Subsidies for agriculture have been significantly reduced. • In the 1998 budget, 26 percent is allocated for agriculture out of which about 75 percent (about $80 million) is direct support to agricultural production and producers • The largest remaining subsidy in 1998, is related to US $33 million dollar credit scheme, where interest rates are subsidized up to 50 percent of commercial interest rates.	• Maintain liberal pricing policies and increase transparency on local markets; • Use domestic market prices for Government purchases; • Focus support programs on efficiency enhancement programs if any allowed by budgetary situation.
B. Trade Policies	• Foreign trade is liberalized but seriously constrained by bureaucratic and informal impediments; • No export tariffs on any agricultural commodities; • VAT on agricultural goods is a strong impediment to exports to other CIS countries (VAT is refunded only for exports to non-CIS countries); • WTO membership is being negotiated;	• Introduce WTO conformed trading regime and complete membership procedures; • Initiate mutual removal of VAT for trade inside CIS; • Phase out Government purchases for strategic grain reserves; • Make any Government purchases of agricultural products on basis of competitive procurement arrangements; • Make strong attempts to sanction interference by local authorities in agricultural markets; • Cease any Government guaranteed machinery and input purchase programs.

ISSUE	STATUS OF REFORMS	OBJECTIVES PROPOSED ACTIONS
C. Taxation.	• Government continues to purchase grain for strategic reserves; level of purchases in 1997 was 700,000 metric tons; • Domestic markets are regularly distorted by the intervention of local authorities; • Machinery markets are distorted by Government guarantees and centrally managed imports of machinery backed by US EXIM Bank. • Incorporated Agricultural enterprises are taxed as other business entities (30 percent profit tax) as of 1998; • Individual private farms pay personal income tax (marginal rates of five to 30 percent), from 1998; • Land tax is paid on all farming land at an average rate of about $20 to $25 per hectare depending on quality of land.	• Maintain an equitable and fair taxation of agriculture; • Consider tax reduction as the most preferable way of supporting emerging family-based private agriculture; • Consider deductions of land-tax and part of the interest on agricultural loans from personal property tax;
2. Land Reform and Farm Restructuring	**Initial Privatization of large farms has been completed, however, the required restructuring of the farming sector is still in progress.** • Land Law passed in 1995 maintained state ownership of agricultural land; • Permanent and freely transferable use rights were provided to members of the state and collective farms on a share basis; • Land for houses and garden plots became private properties; • Farm privatization progressed based on the provision of land use rights and asset ownership certificates; • Most of the farming population decided to remain in the framework of large corporate or cooperative farming enterprises, though a significant and growing number of farm members have exited the farms and established their own family farms; • From 2,500 state and collective farms about 9,000 partnerships, joint stock companies, and cooperatives were created by the middle of 1997; • Individual family farms totaled approximately 60, 000 farming on about 20 percent of agricultural land; • The farm restructuring and ownership changes to date have not yielded increases in farm productivity or improved profitability on most of the farms; • Farm privatization and restructuring has been characterized by a lack of transparency, poor information, and manipulation and has led to a rapid accumulation of land use rights and property shares in the hands of farm managers.	**Efficient farming based on transparent ownership and land use rights.** • Improved transparency and information about ownership rights and opportunities; • Discourage permanent transfer of land-use rights in favor of fixed-term leases, until transparent lease and user-rights markets emerge; • Complete the roll-out of a national land and real estate registration system; • Facilitate the further restructuring of large farms and establishment of additional family farms; • Explore measures to reverse negative social impact of excessive concentration of land user rights; • Facilitate the financial consolidation of newly emerging farms by debt settlement and introduction of international accounting practices; • Use bankruptcy procedures to accelerate the farm consolidation.

ISSUE	STATUS OF REFORMS	OBJECTIVES PROPOSED ACTIONS
3. Competitive Agroprocessing and Services for Agriculture	**Formal privatization of state enterprises completed, technological improvement and financial consolidation of these enterprises is lagging behind.** • The state-owned enterprises were privatized with 50 percent of the shares distributed to the suppliers of raw materials, and 10 percent going to workers in the enterprises. The remaining shares were auctioned; • Capacity in most of the former state enterprises is significantly underutilized, and product quality has only marginally improved; • Increased private sector entry into processing, particularly in the grain sector, creating smaller and more efficient enterprises producing on an international standard quality; • The grain sector is demonopolized with increased strategic investment by large foreign multi-nationals; • Increasing private sector entry into input supply and output marketing; • Current procedures not fully conducive to foreign participation in agroprocessing; • Basic anti-monopoly and fair competition legislation is currently being developed ;	**Private-based and competitive agroprocessing and input supply and facilitation of entry of new private entrepreneurs.** • Enact anti-monopoly and fair competition legislation; • Aggressive application of bankruptcy legislation to facilitate further restructuring of agroprocessing; • Improve the legal and policy environment for direct foreign investment in agroprocessing; • Facilitate the introduction of international quality standards; and • Facilitate the development of secondary markets in shares of privatized agroprocessing industries.
4. Rural Financing	**Basic financial sector reforms in place, with rapid emergence of private financial intermediaries, however the development of a rural financial system is lagging behind.** • Banking system reform complete resulting in the emergence of mainly private and solvent commercial banks; • Lack of collateral, limited experience with rural lending, and poor creditworthiness and the inherited lack of credit discipline of large agro-enterprises limit the amount of rural lending; • With the exception of some pilot projects, individual private farming is not receiving any formal bank financing; • The recent introduction of subsidized credit scheme using budgetary resources is of concern as it is not conducive to the development of commercial rural financing; • Bankruptcy regulations for agricultural enterprises adopted; • Draft legislation on registration of pledges on moveable property is currently under review by Parliament.	**Viable financial institutions efficiently serving the food and agricultural sector** • Improve legal framework for use of land as collateral • Pass legislation on registration of pledges on moveable property; • Introduction of international accounting standards; • Implement Bankruptcy provisions; • Develop and implement legislation on the development of bonded warehouses and the use of warehouse receipts as collateral; • Terminate all programs of subsidized credit for the sector.

ISSUE	STATUS OF REFORMS	OBJECTIVES PROPOSED ACTIONS
5. Institutional Framework	**Adjustment of the institutional framework to meet the requirements of a market-oriented agricultural sector is still at an early stage.** • Government structure in general still reflects Soviet structures, though most of the command economy practices have been abandoned; • The size of Ministry of Agriculture was downsized, but needed structural changes have not been implemented; • Research and Agricultural training and education has not been adjusted and is seriously hampered by budgetary difficulties; • Recent Government reorganization further the fragmentation of research and education; • Western type of extension system does not exist; • The creation of quality public services required by a market based agriculture (market information, animal disease control, phylo-sanitary regulations) is delayed by budgetary limitations.	**Effective provision of public goods and support services.** • Accelerate the creation of quality public administration in agriculture; • Implement comprehensive reform of agricultural research and education; • Facilitate the emergence of a Western-type of advisory service; • Improve provision of public goods such as information, land titling and registration, quality and disease control.

KYRGYZ REPUBLIC

Total Population	4.6 mil	Food and agriculture in GDP (1996)	47%	Agricultural output in 1996 as percentage of 1990 level	70%
Rural Population	65%	Food and agriculture in active labor (1996)	49%		
Total Area	20 mil ha	Food and agriculture in exports (1996)	38%	Livestock production in 1996 as a percentage of 1990 level	
Agriculture area	11 mil ha			milk-------74%	
Pasture	10 mil ha	Food and agriculture in imports (1996)	23%	eggs-------22%	
Arable land	1.4 mil ha			meat-------71%	
Cultivated	83%			wool-------31%	
Pasture	12%	Traditionally an exporter of white sugar, cotton, alcoholic beverages, leather, tobacco		Share of livestock in agriculture (1996)	39%
Forest	5%				

ISSUE	STATUS OF REFORMS	OBJECTIVES PROPOSED ACTIONS
1. Macro-economic Framework for Agriculture A. Prices/Subsidies	**Markets, prices and the trade regime are liberalized; but distortions exist at the local level; market structures are not yet developed, are not competitive and are not integrated.** • Agricultural producer and consumer prices are deregulated. Notable exceptions are irrigation water, electricity and railway tariffs. • Agricultural producer and consumer subsidies are abolished. Notable exceptions are selected remote areas and some agricultural inputs. • Social safety-net in the rural areas is inadequate and not very effective. • Irrigation water and electricity for agricultural use remain subsidized. • Prices for most agricultural outputs are below world prices and prices for most agricultural inputs are at or above world prices. • Input and output markets remain very weak due to poorly developed infrastructure, institutions and information.	**Removal of aany existing distortions in the markets, prices, trade regime and the incentive system; development of fully functioning, competitive and integrated markets for agricultural inputs and outputs.** • Eliminate any interference in the functioning of markets at the local level. • Improve targeting and delivery of social services to the rural poor, particularly in the remote areas. • Increase irrigation water charges gradually to increase cost recovery of O&M cost, or more preferrably, transfer O&M responsibility to users. • Increase electricity tariffs to improve cost recovery and reduce economic losses. • Establish competitive input and output markets, with a primary focus on infrastructure, institutions and information.

95

ISSUES	STATUS OF REFORMS	OBJECTIVES PROPOSED ACTIONS
B. Trade Policies	• Trade regime is generally liberalized but there are still many non-tariff barriers to trade. • WTO accession in progress. • Member of Customs Union (Russia, Belarus, Kazakhstan and the Kyrgyz Republic) but the Union does not seem to be working.	• Eliminate unnecessary and inappropriate non-tariff barriers to trade. • Complete accession to WTO. • Make the Custom Union work to promote free trade.
C. Taxation	• Effective January 1, 1997, several agricultural taxes have been consolidated into one tax, the land tax. • Tax burden (particularly in the agro-industrial sector) is very high but collection rates are generally low.	• Re-assess and revise the land tax rate annually to improve tax revenue and inter-sectoral equity. • Need to rationalize tax rates in order to improve tax revenue, incentives and equity; and improve tax collection.
2. Land Reform and Farm Restructuring	**Significant progress has been made in land privatization and farm restructuring but the process is not yet complete.** • Legal framework has been put in place to establish family farms or restructured large farms with 99-year land use right. • 25% of the arable land is placed in a Land Redistribution Fund (LRF) and 50% of the remaining 75% of arable land is distributed. • The number of state/collective farms has been reduced from 504 to 54 and about 38,000 private farms have been established. • Thus far, land privatization and farm restructuring has been accomplished through many decrees issued by the President or by the government. • Several land laws are now in legislation. • A land registration system has been piloted in two oblasts. • Pasture land, irrigation infrasturcture land and forest land remain state property.	**Completiong of legal framework to develop fully functioning land market and accelerate farm restructuring.** • Prepare, in accordance with market principles, and issue Land Code and Land Mortgage Law. • Issue Law on Land Registration. • Clarify procedures for registering rights and transactions in land shares and prepare standard form of contract for sale and lease of land shares. • Issue "regular" land share certificates to those who do not yet have them. • Design an overall framework for auctioning LRF land to ensure efficiency, equity and transparency. • Renew efforts to amend the Constitution to permit private ownership of land. • Examine the use of rights issue for common property such as pasture land, irrigation infrastructure land and forest land. • Educate public in the meaning of the individual rights granted by the land and agrarian reform program.

ISSUES	STATUS OF REFORMS	OBJECTIVES PROPOSED ACTIONS
3. Competitive Agroprocessing and Services for Agriculture	Substantial progress has been made in privatizing the agro-processing and input supply enterprises but the process is not yet complete and the privatized enterprises are not very efficient. • Most of the small and medium scale state owned agro-processing enterprises have been privatized. • The large agro-industrial enterprises are being privatized through case-by-case method of privatization but the process is slow. • The newly privatized enterprises are hampered by obsolete equipment/technology, lack of credit, poor management, and poor understanding of the competitive markets. • Product quality is very poor; and the existing product quality grades and safety standards (which are not being enforced) are not comparable to international grades and standards. • Lack of critical agricultural inputs is a key constraint to increase agricultural productivity. • Foreign direct investment in the agro-industry remains very low.	Completion of privatization of the existing agro-processing and input supply enterprises and the closure of inefficient (those that cannot be made profitable through restructuring) state owned enterprises that cannot be privatized. • Complete the process of privatization of all the existing state owned agro-industrial enterprises. • Simplify registration procedures and reduce the permissions required to establish agro-industrial joint ventures. • Provide increased access to commercial credit (for working capital and capital investment) by increasing KAFC's (Kyrgyz Agricultural Finance Corporation) lending capacity; and management training for enterprise managers. • Improve product quality and packaging through technological improvements; and establish product quality grades and safety standards that are comparable to international grades/standards. • Establish a competitive agricultural input marketing system through the development of private input dealers. • Improve the regulatory environment and economic incentives to promote foreign direct investment in agro-industry.

ISSUES	STATUS OF REFORMS	OBJECTIVES PROPOSED ACTIONS
4. <u>Rural Financing</u>	The availability of credit is a serious constraint to rural development. However, initial steps to establish a commercial rural credit system have been taken. • Agroprombank has been liquidated. • There are very few commercial banks and none is interested in lending to agriculture. • Kyrgyz Agricultural Finance Corporation (KAFC) -- a non-banking, independent and commercial public financial institution -- has been established. • Interest rates have been increased to positive levels in 1997. • Outstanding agricultural debts are being recovered	Acceleration in the development of viable financial institutions serving the rural and agricultural sectors. • Accelerate the implementation of Rural Finance Project (World Bank) and Rural Agriculture Development Bank (ADB). • Phase out budgetary transfer for agricultural credit in 1999. • Interest rates charged on agricultural credit through budgetary transfers should not be lower than those charged by the KAFC. • Resolve the issue of outstanding agricultural debt by June 2000. • Support the establishment of "dealer credit" and "trade finance" and expand the lending operations of KAFC to provide credit to meet working capital and capital investment needs of agriculture and agro-industry.
5. <u>Institutional Framework</u>	The institutions of "planned" economy are gradually being replaced by institutions that serve private agriculture based on market principles. However, the process is very slow. • The Ministry of Agriculture and Water Resources is being reorganized. • The agricultural research, extension and education systems have not yet been adjusted to the emerging market conditions. • Information system required for market-based private agriculture is not yet in place. • The institutional capacity to undertake agricultural policy analysis is very limited.	The process of establishing efficient and effective institutions to serve commercial private agriculture and rural sector needs to be accelerated. • Complete the reorganization of the Ministry of Agriculture and Water Resources to serve as the main agency to implement agricultural policy. • Strengthen agricultural research, extension and education systems to serve the needs of commercial agriculture in the private sector. • Establish market and technical information systems to collect, process and disseminate appropriate information to emerging private farmers. • Strengthen institutional capacity and training of staff involved in designing and implementing agricultural and rural development programs, including agricultural policy analysis.

MOLDOVA

Total Population	4.4 mil	Food and agriculture in GDP (1996)	40%	Agricultural output in 1996 as percentage of 1986-1990 level		45%
Rural Population	52.5%					
		Food and agriculture in active labor (1996)	50%	Livestock production in 1995 as percentage of 1986-1990 level		44%
Total Area	3.4 mil ha					
Agriculture area:	2.3 mil ha	Food and agriculture in exports (1992)	55%	Share of livestock in agriculture. (1995)		31%
				Agricultural area in private use (1997)		20%
Arable land	76%	in imports (1992)	14%	Share of independent private farms in total agricultural land (1996)		6%
Orchards	14%	Traditionally net exporter: wine, processed and unprocessed fruits and vegetables, and pork.				
Irrigated	10%			Share of private sector in total agricultural output (1996)		40%
Forested	12%					

ISSUE	STATUS OF REFORMS	OBJECTIVES PROPOSED ACTIONS
1. Macro-economic Framework for Agriculture A. Prices/Subsidies	**Significant progress in price and market liberalization, with recent setbacks.** • Most producer prices were liberalized in 1993-95. • State order for grain delivery for state reserves was re-introduced for the 1996 and 1997 cropping season in the form of tax payments and social security contributions payable to the state in the form of grain. • Farms with debt to budget received only 50% of USD 160-180/ton grain price in 1996. • Bread prices were liberalized in late 1996. • Milk producer prices are kept low with informal interventions and by subsidies provided in the form of tax deductions. • Consumer prices are liberalized except dairy products. • Direct producer subsidies are mainly phased out. • Agriculture receives indirect subsidies via special energy tariffs and input supply schemes.	**Minimal Government intervention in agricultural markets.** • Return to the liberal course of market and price policies. • Remove remaining interventions from grain and milk markets. • Phase out indirect subsidies to agriculture. • Deregulate fully the processing and trade margins. • Re-monetization of the sector by proper treatment of inter-enterprise arrears, accumulated debt and revision of tax and social security payment policies.

ISSUE	STATUS OF REFORMS	OBJECTIVE PROPOSED ACTIONS
B. Trade Policies	• Domestic trade is liberalized, but a competitive market structure is not in place. • Government procurement is restricted to grain, purchased for state reserves. • Export of agricultural products, including grains was liberalized in 1997. • Non CIS imports are subject to 10-30% but occasionally higher (50%) tariffs.	• Establishment of commodity exchange or promotion of the use of commodity exchanges in other countries. • Introduction of market methods in procuring grain for government purposes and for marketing of grain and input deliveries based on inter-governmental agreements. • Introduce low and uniform tariffs and maintain a policy of no quantitative restrictions on exports. • Develop a strategic grain stock policy and trade policy.
C. Taxation	• Agriculture pays land tax, social tax. • About 40% of taxes are collected. • Private farms almost fully fulfill their tax obligations. • Agriculture enjoys some VAT preferences for imported inputs, the system, however, needs to be upgraded.	• Develop a fair and non-discriminatory system of agriculture taxation. • Improve tax collection rates.

ISSUE	STATUS OF REFORMS	OBJECTIVE PROPOSED ACTIONS
2. Land Reform and Farm Restructuring	**Slow process of farm restructuring and privatization.** • Additional land was provided for household plots and for gardening. (In 1995, 13% of agriculture area.) • Land shares were allocated for the farming population and 91% of beneficiaries received certificate. • Number of independent private farms are increasing fast but their share in land use is still small. (1997, 7% of agriculture area.) • Pilot projects (Nisoporeii, Orhei) are progressing well and provide experience for a national program of farm restructuring, which is progressing well. • Farm registration and exit procedures were simplified in early 1997. • About 120,000 private farms and over 200 associations involve about 180,000 beneficiaries of land reform out of a total of 983,000. • Most of the large-scale farms are operating without any significant restructuring and are accumulating significant debt. However, the break up and restructuring of larger farms is accelerating. • February 1995 Amendment of Land Code allowing only group exits from large farm was eliminated by the Constitutional Court in early 1996. • The moratorium of agricultural land sales, up to 2001 was declared unconstitutional by the constitutional court. Sales of land certificates inside the large farms and sales of urban and agricultural land is allowed. Procedures for agricultural land sales are being prepared. • The emergence of land markets is seriously constrained by the absence of a well functioning land titling and registration system. • Significant steps were made to establish a uniform cadastre system in 1997 and 1998.	**Farming structure based on secure transferable land use rights.** • Facilitate the process of farm restructuring and improve the process of withdrawal of land and non-land assets from collective farms. • Make the establishment of new private farms, and farming enterprises based on partnership or cooperation as easy as possible. • Use reserve land whenever possible for providing land for new beneficiaries or offer financial compensation instead of decreasing already allocated land shares. • Market mechanism for rent, leasing and the establishment of collateral should be developed as soon as possible with the aim of facilitating land consolidation and financing efficient farming. • Adopt and implement a program of debt settlement for large scale farms. • Accelerate the process of the creation of a suitable property and land registration system.

ISSUE	STATUS OF REFORMS	OBJECTIVE PROPOSED ACTIONS
3. Competitive Agroprocessing and Services for Agriculture	**Privatization and demonopolization is in progress.** • Agroprocessing and input supply industries are privatized as part of the overall privatization framework. • Producers of agricultural raw materials received 50% of ownership of agroprocessing. • The initial privatization of agro-processing was completed by the end of 1995. The majority of enterprises, however, are unconsolidated and lack transparent ownership. • Significant portions of shares in agroprocessing industries have not been sold and still belong to the state. • Most the agroprocessing sector enterprises are still operating the old fashioned way and are in effect bankrupt. • Cerealea (grain corporation) and Fertilitatea (input supply corporation) are in the process of privatization after initial demonopolization but the process is not fully completed. • Anti-monopoly regulations are not fully in place. • Increasing but still limited entry of private sector into processing and input supply. • Limited, but growing, foreign participation in privatizing agroprocessing.	**Private competitive processing and input supply industries.** • Increase role of auctions in the privatization process. • The shares of companies based on privatized patrimonial bonds should be openly traded as soon as possible. • Promotion of the participation of foreign investors both in auctions and in the second round of privatization. • Introduce post-privatization programs to facilitate the restructuring of privatized enterprises. • Develop and implement effective anti-monopoly legislation. • Strictly enforce bankruptcy legislation in order to consolidate the newly established private sector. • Complete the Privatization of Cerealea and Fertilitatea together with the appropriate restructuring and real privatization of Moldcoop.
4. Rural Financing	**Lack of an appropriate financial system for food and agriculture.** • Financing in agriculture is not adjusted to the needs of a market based privatized agriculture. • High interest rates and the lack of collateral seriously limits lending to agriculture. • Development of a rural credit system is underway; with World Bank support the first 200 rural credit cooperatives have been established.	**Viable financial institutions efficiently serving the food and agriculture sector.** • Introduce processors or suppliers' credit schemes to finance farming inputs. • Prepare an action plan to revitalize and financial services. • Implement pilot projects to establish credit unions in the villages.

ISSUE	STATUS OF REFORMS	OBJECTIVE PROPOSED ACTIONS
5. Institutional Framework	**Adjustment of the institutional framework is at a rather early stage.** • Government administration reflects Soviet-type structures. • Research/education system has not been adjusted to emerging new conditions. • Public activities (government research-education) in agriculture are seriously hampered by budgetary difficulties. • Western type of agricultural extension system does not exist. • A non-Governmental Agricultural Restructuring Agency (ARA) was created to support farm restructuring.	**Efficient and effective public sector administration and support services.** • Re-orient Government attention toward private agriculture. • Prepare and implement a program of re-organization of public administration in agriculture. • Reorganize the Ministry of Agriculture and Food. • Review the agricultural education and research system. • Support the emergence of private farm advisory services for landowners.

RUSSIA

Total Population	148 million	Food and agriculture in GDP 1995	8.4%	Agricultural output in 1996 as percentage of 1989 levels	63%
Rural Population	27%	Food and agriculture in active labor (1995)	15.7%	Share of livestock in agriculture (1995)	48%
Total Area	1708 mil ha	Food and agriculture in exports (1996)	4.7%	Agricultural area in private (non state and non cooperative) use (1996)	8.6%
Agriculture area:	222 mil ha	in imports (1996)	25.3%	Share of family private farms and household plots in total agricultural output (1996)	48%
Arable land	38%	Net importer of livestock and dairy products and sugar			
Pasture and Meadows	40%				
Forested	45%				

ISSUE	STATUS OF REFORMS	OBJECTIVES PROPOSED ACTIONS
1. Macro-economic Framework for Agriculture	**Substantial liberalization at the federal level. Less liberal policies prevail at the regional level. Inflation expected in single digits in 1998.**	**Consolidation and maintenance of macroeconomic stability. Removal of barriers to trade at the subnational level, reduction of intervention at the provincial level, integration of national markets and complete linkage with global markets.**
A. Prices/Subsidies	• Little price intervention at the federal level. Substantial intervention in the form of barriers to inter-regional trade and state procurement by provincial governments in selected provinces. • Subsidies and support for agriculture in 1996 reduced to 2.4% of federal budget, and 4.9% of regional budgets, or 3.9% of consolidated budget, with further reduction in 1997.	• Enforce legal prohibition on interference in inter-regional trade; require competitive procurement by regional governments • Within sectoral budget consistent with macroeconomic stabilization, achieve expenditure switching toward public goods and services, and away from programs that distort incentives.
B. Trade Policies	• Current tariff regime is liberal and tariffs are low (in the range of 15%) and relatively uniform (except sugar at 25%). Pressure to increase protection is strong, especially for poultry and alcoholic beverages. Government expects accession to WTO in 1998.	• Continue to make substantial progress toward joining WTO. Maintain current low and uniform tariff structure. Do not introduce quantitative controls inconsistent with rules of WTO.
C. Taxation	• New tax code approved in first reading in 1997, continued negotiation 1998. Code limits tax exemptions and rights of subnational jurisdictions to impose ad hoc taxes. Under new code agricultural producers would find it harder to evade taxes, but would be subject to fewer and less distorting taxes. Private farmers have five year holiday on land taxes.	• Evaluate agricultural provisions of new tax code. Monitor enforcement of restrictions on discriminatory ad hoc local taxes.

ISSUE	STATUS OF REFORMS	OBJECTIVES PROPOSED ACTIONS
2. Land Reform and Farm Restructuring	**Significant progress in the privatization of agricultural land; state retains ownership of approximately 10%. Most of remaining land owned in shares by agricultural enterprises and their members/owners. Land tenure in latter is ambiguous, since rights of enterprises and shareholders not fully specified through contracts governing land use.** • Distribution of land to enterprises and their shareholders undertaken through a series of Presidential decrees since 1992. • Private land ownership recognized and protected by law. Purchase and sale of agricultural land legal under provisions of Presidential decree, but not yet guaranteed by passage of a land code that will take precedence over presidential decrees. • Land markets are largely inactive due to legal ambiguity regarding transactions and lack of clarity on procedures. • In 1997 there were 279,000 private farms, holding about 4.8% of available agricultural land, household plots accounting for 4.5% of agricultural land. Agricultural enterprises of various forms cultivated about 90% of agricultural land.	**Removal of remaining obstacles to development of a private land market.** • Evaluate state farms remaining on list of unprivatized. Reduce number remaining in public sector. Complete drawing up of contracts between shareholders and users of land (enterprises). Include registration of land shares in system of land registration. • Speed up land survey registration and titling service as a priority to developing a fully functional land market. • Establish and disseminate procedures for land transactions, including leasing as well as purchase and sale. • Establish a framework for the dissemination of land market information. • Establish implementable procedures for restructuring and/or liquidation of insolvent farm enterprises, so that resulting enterprises have incentives and capacity to engage in commercial activities.
3. Competitive Agro-processing and Services for Agriculture.	**Considerable progress in privatization, but investment remains low, transactions costs are high, and much of the processing industry is not competitive with regard to price and quality.** • Foreign participation in marketing and agroprocessing is lower than desirable. • Interventionist activities of local and provincial governments constrain marketing opportunities for agroprocessors and discourage investment. • Development of information system and marketing infrastructure in the food chain is inadequate.	**Removal of legal and institutional constraints to market development.** • Enforce bankruptcy provisions in food processing. Enforce prohibitions on restrictions of inter-regional trade. Tie federal assistance on agricultural support programs to reduced procurement by provincial governments. Evaluate structure of rail rates. • Create farmer/processor market information systems. • Expand and upgrade public facilities for output marketing (farmers' markets, wholesale markets, etc.). • Improve collection and dissemination of market information.

ISSUE	STATUS OF REFORMS	OBJECTIVES PROPOSED ACTIONS
4. Rural Financing	**Agroprombank has merged with a strong private bank, SBS, to form SBS-Agro. Conversion of activities and portfolio of the former Agroprombank to fully commercial basis is underway.** • Real interest rates remain high and impede borrowing by the agricultural sector. • Suppliers' credits becoming a more prevalent form of finance in the rural sector	**Development of multiple channels for rural finance.** • Identify and remedy deficiencies in the legal framework for collateralizing moveable property. • Design and implement a program to facilitate expansion of leasing of agricultural equipment by the private sector, with a sunset clause for termination of public support for the activity • Support activities to reduce the transactions costs that rural lenders and borrowers face, such as information and evaluation of risk
5. Institutional Reform	**Institutional reforms are uneven, with excessive local intervention** • Access to information and advisory services insufficient to support competitive production and marketing • Agricultural research is underfunded and insufficiently linked with international partners • Local administrations retain an excessive role in market intervention • Agricultural education has not been revitalized, and remains under the jurisdiction of the Ministry of Agriculture and Food	**Reduce local involvement in procurement and refocus research and education** • Improvement in market information systems is underway, and needs strengthening • Increased funding and reorganization of agricultural research is needed to focus on increased competitiveness and international integration • Local administrations should reduce procurement into regional funds and conduct remaining procurement on competitive basis • Agricultural education should be refocused to address rural labor mobility more generally, and should be managed under the Ministry of Education

TAJIKISTAN

Total Population	5.8 million	Food and agriculture in GDP 1996	26.6 %	Agricultural output in 1995 as percentage of 1990 level	65%
Rural Population	67%	Food and agriculture in active labor (1996)	50 %	Livestock production in 1995 as percentage of 1990 level	46%
Percapita Income (1996)	US $330	Food and agriculture in exports (1996)	32%	Share of livestock in agriculture (1995)	36%
Total Area		in imports (1996)	22%	Arable area in private use (1995)	9 %
Agriculture area:	14 mil ha	Traditionally a major exporter of cotton and in addition processed and unprocessed fruits, vegetables, nuts, silk, and wine, and a net importer of: grain, sugar and vegetable oil,			
Arable land	1 mil ha				
Orchards,	7%				
Vineyards	8%				
Irrigated	78%				

ISSUE	STATUS OF REFORMS	OBJECTIVES / PROPOSED ACTIONS
1. Macro-economic Framework for Agriculture A. Prices/Subsidies	Following years of political instability and civil war, the recent peace accord and renewed government commitment to economic reforms and stabilization provide favorable environment to develop the private sector role in the agricultural sector, which is in its infancy. • Prices of fruits, vegetables and livestock products and inputs fully liberalized; • Cotton and wheat prices partially liberalized; Cotton export prices are based on Liverpool prices and indexation of local currency price to exchange rate depreciation. Grain prices were close to international prices during 1996; • Irrigation water continues to be significantly subsidized • Bread prices fully liberalized, universal bread subsidy replaced with targeted food subsidies to the most vulnerable groups	Creation of an enabling environment for private sector participation; development and implementation competitive and fair agricultural markets without Government intervention, and improved targeting of support programs. • Continuation of macro-economic adjustment operations and capacity building to implement market oriented agricultural policies; such as liberalization of all prices and phasing out producer subsidies. • Establish competitive and functioning agricultural markets in the private sector; • Provide secure trade routes from farm to markets and eliminate border restrictions. • Monitor existing support programs and improve targeting to the absolute poor. • Continue BOP support to provide adequate foreign exchange for import of agricultural inputs using the private sector

ISSUE	STATUS OF REFORMS	OBJECTIVES PROPOSED ACTIONS
B. Trade Policies	• privatization cotton ginneries nearly completed • Elimination of the state order system for all agricultural commodities • Licensing requirements for the import of agricultural inputs and export of all agricultural exports, except cotton, tobacco and silk have been abolished • Government continues to impose reserve requirement of 20%	• Abolish domestic reserve requirements • Monitor progress of cotton stock exchange and expand linkage to regional and world markets and promote conditions for active forward trading activities for cotton. • Remove remaining export licensing and quotas • Market for cotton through "Tojakpakthe" an association of cotton traders and processors and producers.
C. Taxation	• Present laws impose several types of taxes on the agricultural sector: e.g. production tax, land tax, irrigation tax, transport tax; and are subject to varied interpretation; • Registration a requirement for all those engaged in marketing of agricultural products • Cotton exports taxed at 25% of FOB price	• Rationalize the taxation system, • Strengthen the Government capacity in formulation and implementation of appropriate tax policies • Phased reduction of cotton sales tax • Monitor the government policies of the reduction of all export taxes and duties for agricultural products
2. Land Reform and Farm Restructuring	Land privatization and farm restructuring are taking place in an ad-hoc manner and are at early stages • The Tajik Parliament has passed several land laws, decrees and distributed 50,000 ha for the creation of peasant (dehkan) farms according to which more than 240,000 citizens received from 0,08 to 0,15 hectares of irrigated land and up to 0,50 hectares of un-irrigated land for inheritable life long use for agriculture October 1995. • Government has provided long-term lease rights for the dehkan farms as a first step towards land privatization of former state and collective farms • Most state and collective farms have been converted into joint stock companies or associations, but without major change in the mode of operation • Degree of commitment to land privatization and farm restructuring varies across the country depending on the hukumat • Secure land registration and titling services and supporting mechanisms do not exist • Majority of ex-state and collective farms are in serious financial condition, • Almost all the state farms, except the one dealing with pedigree breeds and seed multiplication, have been transformed into collective agricultural enterprise without any significant change in their structure, mode of control and operations.	Development of transparent, participatory approaches for equitable distribution of land, creation of fair and competitive land lease markets • Improve the legal framework for land reform through (a) allowing land use rights for legal persons (individuals, true co-ops and corporations); security of land tenure with rights to exit, and formalizing clear rights to own and sell land; (b) constituting a participatory and transparent mechanism for determination of land and non-land assets for individuals and their allocation; (c) mechanisms to use land lease rights/ other assets as collateral • Modernize the land registration systems and titling services in the immediate to short term to develop a functioning land lease market • Reform the legal framework for individuals, cooperatives and corporate entities in agriculture to provide for transparency, autonomy and framework for easy restructuring of farms and agencies and enterprises providing farm support services. • Develop a variety of private farming approaches and provide legal and appropriate institutional support services for their creation and sustainability. • Rehabilitate critical irrigation and drainage infrastructure and reduce reliance on pumped systems, except where no alternatives exist; and develop rainfed agriculture

ISSUE	STATUS OF REFORMS	OBJECTIVES PROPOSED ACTIONS
3. Competitive Agro-processing and Services for Agriculture.	New privatization law passed by parliament (May 16, 1997) and signed by President (June 11, 1997), and new corporatization procedures expected to improve the legal framework and privatization of agro-processing sector • Government continues to retain partial to majority ownership in most agro-enterprises • Foreign participation in marketing and agroprocessing is minimal • Potential of state and collective farms becoming major shareholders of food processing enterprises • High level of indebtedness, low capacity utilization and low quality products major impediments	Expansion of private ownership, formulation and implementation of transparent legal and privatization procedures for the development of a fair, competitive agroprocessing and input supply markets. • Develop and implement a plan for complete privatization of all agro-processing and input service enterprises, undertake case by case privatization, with participation of both domestic and foreign investors. • Create enabling policy environment to attract private foreign investment; improve legal system for contract enforcement and market transparency. • Develop and implement anti-monopoly legislation. Implement monitoring and regulatory mechanisms of the privatization process to prevent oligopsony comprising both domestic and foreign cotton ginnery owners • Promote research and development of new products, packaging and marketing to meet outside markets;
4. Rural Financing	A functioning broad based rural financial system is at a rudimentary stage • Farms and agro- processing agencies/enterprises are in serious liquidity crisis • High interest rates and the lack of collateral seriously limit lending to agriculture • Financing through Agroprombank has declined substantially, • Micro credit schemes are being implemented in selective regions using NGO's • National system of rural credit cooperatives was created with EU. • Mortgage Bank is under consideration	Creation of viable market oriented financial institutions to serve the agricultural sector • Restructure Agroprombank into an autonomous commercially viable rural financial institution and train loan officers in market oriented agricultural lending ; implement accepted accounting principles and prudent banking practices; • Support the establishment of a variety of rural credit delivery mechanisms • Develop and implement mechanisms to promote private input/output marketing and trading services

ISSUE	STATUS OF REFORMS	OBJECTIVES PROPOSED ACTIONS
5. <u>Institutional Reform</u>	**Reforms to restructure government institutions are in early stages.** • Ministry of Agriculture continues to operate in the Soviet type Government structure. • Agricultural Research, Extension and Education system have not been adjusted to emerging market conditions. • Reorganization of the research system is planned but, due to serious budgetary difficulties, cannot be implemented for the near future. • Information system required by a market based agriculture is not in place.	**Redefinition of the roles of public and private sector institutions to support competitive, market oriented agriculture sector.** • Complete the reorganization and improve quality of public agricultural administration to the needs of a market economy. • Complete the reform of agricultural education and research. • Establish public information system to provide better understanding of rights of individuals and voluntary groups regarding the process of determination and allocation of shares at the grass roots level. • Develop a strategy and establish an institutional mechanism for implementation of a multi-stage system of management for the rational basis for land use rights distribution by defining the roles, functions and authority of the State Land Committee, Ministry of Agriculture, Ministry of Justice, Agrarian Reform Commissions at various levels.

TURKMENISTAN

Total Population	4.3 million	Food and agriculture in GDP 1995	25%	Agricultural output in 1996 as percentage of 1989 levels	60%
Rural Population	2.5 million	Food and agriculture in active labor force (1995)	44%	Share of livestock in agriculture (1995)	40%
Total Area	49.4 million ha.	Food and agriculture in export (1996)	20%	Agricultural area in private (non-state and non cooperative) use (1996)	10%
Agriculture area:	40.7 million ha.	in import (1996)	20%	Share of private farms and household plots in total agricultural output (1996)	25%
Arable land	3%				
Pasture and					
Meadows	81%	Net exporter of fiber, importer			
Forested	4%	of food			

ISSUE	STATUS OF REFORMS	OBJECTIVES PROPOSED ACTIONS
1. Macro-economic Framework for Agriculture A. Prices/Subsidies	**High levels of government intervention in agriculture including state orders for the two major crops, wheat and cotton, subsidies for agricultural inputs, direct control of trade, and taxation through the State Order System.** • Measured inflation is relatively low (under 20% annually) but suppressed inflation may imply a higher actual rate. • Foreign exchange controls remain in effect, and the manat is over valued at the official exchange rate • Arrears in collection of payments owed for energy exports cause severe disruption throughout the economy. • Main products of the crop sector, cotton and wheat, remain fully subject to state orders at prices which are approximately half of international trading prices. • Inputs subsidized by 50% for production of cotton and wheat. • No significant payments for irrigation water	**Phase out Government control of agricultural markets.** • Adopt program of macroeconomic stabilization proposed by IMF. • Adopt program to phase out state orders over two year period. • Phase out input subsides over same period. • Introduce water fees for partial cost recovery of irrigation costs.
B. Trade Policies	• Imports and exports registered through state commodity exchange, which amounts to defacto licensing requirement • Government controls all cotton exports • Minimum export prices for hides and skins.	• Eliminate registration requirement for imports and exports. • Develop a free market for cotton exports • Eliminate minimum prices on hides and skins.

ISSUE	STATUS OF REFORMS	OBJECTIVES PROPOSED ACTIONS
C. Taxation	• Implicit taxation high through state order system and export controls, but little explicit taxation. • Private owners pay a land tax, but collection sporadic.	• Phase out implicit taxation through state orders and export controls. • Gradually increase land tax as state orders reduced. • Use land tax to pay for rural social services eliminating mandatory payments to farm associations now providing social services.
2. Land Reform and Farm Restructuring	**Land Reform is Slow but Making Progress** • Land in present associations allocated to households on ten year leaseholds. • About 5% of land in individual private farms. • Leaseholds convertible to ownership upon successful performance for two years.	**Increase the Pace of Land Reform and Private Farmer Support.** • Adopt realistic indicators of successful performance so that at least 50% of leaseholders receive ownership after two years. • Adopt complementary reforms so that private owners can function in market environment.
3. Competitive Agroprocessing and Services for Agriculture	**State Control of All Agroprocessing and Services** • Most processing and services handled by state enterprises organized into large associations. • Very little privatization.	**Adopt and Implement Program of Privatization and Demonopolization.** • Privatize and demonopolize existing agribusiness companies and associations. • Remove barriers to new entry in processing and services.
4. Rural Financing	**High Dependence on Subsidized and Directed Credit through Government Controlled Banks.** • High Dependence on Subsidized Credit • Several Banks serve agriculture and the dominant Bank is Daikhan Bank. Daikhan Bank has little risky debt from the past, but is not yet engaged in genuine intermediation.	**Reduce Subsidized Credit and Introduce Commercial Banking Practices** • Reduce subsidy element of directed credit. • Conduct financial audit and develop corporate plan for Dankhan Bank.

ISSUE	STATUS OF REFORMS	OBJECTIVES PROPOSED ACTIONS
5. Institutional Framework	**Non-transparent Budgeting and Inadequate Resource Allocations to Public Institutions**	**Consolidate Government Budget and Increase Expenditures for Public Goods**
	• Extra budgetary Agricultural Development Fund manages sectoral financial flows in non-transparent fashion.	• Consolidate costs and revenues from sector into general budget and abolish Agricultural Development Fund.
	• Frequent institutional reorganization and high turnover of administrative staff in response to declining sectoral performance.	• Remove authority of local administration to interfere in decisions regarding production and marketing.
	• Frequent local government interference in production decisions	• As resources permit, develop agricultural research and extension to serve needs of agents in a market economy.
	• Research and extension much reduced	

UKRAINE

Total Population	51 mil	Food and agriculture in NMP 1996	21%	Agricultural output in 1996 as percentage of 1990 level	59%
Rural Population	32%				
		Food and agriculture in active labor (1996)	24%	Livestock production in 1996 as percentage of 1986-1990 level	52%
Total Area	60 mil ha.				
Agriculture area:	42 mil ha.	Food and agriculture in exports (1996)	21%	Share of livestock in agriculture (1996)	43%
				Agricultural area in private use (1996)	24%
		in imports (1996)	8%	Share of independent private farms in total agricultural land (1995)	2%
Arable land	79.3%	Traditionally net exporter: grain, oil seeds, sugar, dairy products, beef and pork.			
Orchards	1.9%				
Irrigated	6.0%			Share of private sector in total agricultural output (1994)	53%
Forested	15.6%				

ISSUE	STATUS OF REFORMS	OBJECTIVES PROPOSED ACTIONS
1. Macro-economic Framework for Agriculture A. Prices/Subsidies	**Significant progress in price and market liberalization constrained by underdeveloped market structures.** • Fixed producer prices were abolished for agricultural commodities in 1994. • Wheat prices were deregulated in early 1995 and bread prices were been increased to reflect harvest costs and subsequent inflation. • Controls on profit and trade margins for wheat, flour and bread products were removed in May 1996. • Lack of competitive domestic markets and underdeveloped trading system still keep producer prices under border prices. • Scale of credit and other subsidies (inputs) substantially reduced in 1995 and 1996 and capped at 1996 levels. • Subsidies include mainly subsidized credit advanced by Government only for spring operations related to grain contracted for sale to the Government.	**Minimal Government intervention in agricultural markets.** • Abstain from implementing new distortive price control measure. • Cap subsidies associated with input prices, support prices, and other producer price subsidies and devise a phasing out schedule for these subsidies in 1997-1998. • Remove implicit taxation on agricultural producers, processors, and marketing firms by promoting competition and domestic market development. • The agricultural credit subsidies capped at 1996 levels are to be brought down to real terms in the future.

ISSUE	STATUS OF REFORMS	OBJECTIVES PROPOSED ACTIONS
B. Trade Policies	• State agricultural purchases reduced in 1994 to roughly 40% of 1989-92 average levels of shares of state purchases in total marketed production. In 1996, these purchases fell to 10% of 1989-1992 average levels. • Total annual government expenditures in the agricultural sector have been capped at about $600 - $700 million. • In 1997 a government guaranteed program to provide fertilizer and feed was introduced, together with a machinery leasing system as part of the budget expenditures • Increased intervention of regional authorities creates new distortions. • Remaining agricultural commodities removed from the export registry system in March 1995. • All agricultural export quotas removed. Remaining quotas on grain exports were removed in February 1996. • Quotas on imports of some meat products were introduced in 1997 • No licensing of agricultural imports and there are average 15% tariffs and VAT on imports • Agricultural imports are taxed on the basis of reference prices	• Create conditions for remonetizing commodity and payment relations. • Ensure that the Customs Service would not impose indicative prices on products not subject to registry. • Ensure that all state agricultural procurements are executed on a competitive basis through the agricultural commodity exchanges, public tenders, and auctions. • Remove export taxes on live cattle and animal skins. • Refrain from intervening in agricultural import and export markets, with the exception of interventions acceptable under the GATT/WTO. • Acquire the capacity to use the safeguards and antidumping measures available under the GATT/WTO. • Pursue active trade policy to improve market access for Ukrainian food and agricultural products through the WTO and Cairns Group, and by seeking duty-free access for these products to the markets of the states of the FSU.
C. Taxation	• 30% profit tax is applied for food industry, while primary agriculture is exempt. • Taxes on intermediaries reduced from 70% to 45% in early 1995, and to 30% in May 1996.	• Promote a fair and non-discriminatory system of taxation. • Reduce differentiation in tax rates across sectors to minimize distortive effects on resource allocation.

ISSUE	STATUS OF REFORMS	OBJECTIVES PROPOSED ACTIONS
2. <u>Land Reform and</u> <u>Farm Restructuring</u>	**Early stage of land reform and farm restructuring moving toward privatized farming.** • Additional land was provided for household plots and for gardening. In 1995 they covered about 5 million hectares, about 11% of agriculture land. • In 1992 10% of agricultural area was set aside for reserve to be used to establish new private farms. • Number of independent private farms are increasing, however, in late-1996, there were only 33.040 covering 1.8% of total agricultural area. • Regulations allow the division of collective and state farmland into physically identified privately owned shares, but procedures are not conducive enough to support the real restructuring of large-scale farms. • Withdrawal of land and asset shares from newly formed joined stock societies is constrained by complicated procedures. • There is a six year moratorium on sales of land obtained from collective farms. • Land registration and titling practice do not meet the needs of a functioning land market. • By March 1997 issuance of land share certificates had been completed to all entitled beneficiaries in 40% of former collective farms. • Simple and transparent procedures for the exchange of property and land shares (or a combination of the two types) for land plots and physical assets were adopted in April 1996.	**Secure transferable land use rights conducive to promoting long term investment, access to financial markets, and enhanced land mobility.** • Abolish the moratorium on sale of land or land shares (currently six years) and modify the current restriction which stipulates that agricultural land which is transferred to new private owners is restricted for agricultural use only. • Clarify that the priority right of the collective farm to purchase the land of a member who has withdrawn and wants to sell his land is a right of first refusal only. • Develop mortgage procedures for land, other real estate, and moveable assets. The mortgage law would allow leasees to mortgage their leasehold interest. • Adopt a resolution to establish a single registry of land and other real estate. • Withdrawal of land and other property from collective and state farm enterprises would not be subject to approval by either farm management or the farm members' council. • Ensure that no legislation or administrative regulation restricts the right to or terms of land leases or sharecropping contracts.

ISSUE	STATUS OF REFORMS	OBJECTIVES PROPOSED ACTIONS
3. Competitive Agroprocessing and Services for Agriculture	**Privatization and demonopolization slow in process.** • State owned agroprocessing and input supply system is in the process of privatization (roughly 3,600 enterprises privatized by March 1998). • The period of the closed subscription phase (under the AIS Privatization Law of 1993) was reduced to a period comparable to the closed subscription period for non-agricultural enterprises. • Over 100 large state-owned agricultural input supply, output marketing, and agro-processing monopolies were demonopolized to create independent enterprises. • Increasing entry of private sector into input supply and services. • Basic anti-monopoly regulations developed during 1993-1994. • Producers of agricultural raw materials have preferential rights to obtain 51% of shares of agroprocessing enterprises. • Ban on privatization of grain procurement and bread industry enterprises was removed in 1995. • Current procedures for privatization are not conducive to foreign investment.	**Facilitation of the emergence of new and privatized restructured firms, and the growth of efficiency in both input and output markets and in agroprocessing.** • Accelerate the privatization of grain procurement and bread industry. • Create a policy and legal environment supportive of direct foreign investment.
4. Rural Financing	**Lack of an appropriate financial system for food and agriculture.** • Financing in agriculture is not adjusted to the needs of a market based privatized agriculture. • Commercial banks exist, but do not lend to agriculture. • High interest rates and the lack of collateral seriously limits lending to agriculture. • Budgetary allocations and credit emissions for agricultural procurements were reduced significantly in 1995-96 under the reform programs supported by the IMF and World Bank stabilization and structural adjustment operations.	**Viable financial institutions efficiently serving the food and agriculture sector.** • Phase out all credit subsidies. • Assess the current structure of rural finance and recommend methods to defray transaction costs and reduce risk in rural lending. • Eliminate distortionary credit policies.

ISSUE	STATUS OF REFORMS	OBJECTIVES PROPOSED ACTIONS
5. Institutional Framework	Adjustment of the institutional framework is at a rather early stage.	Efficient and effective public sector administration and support services.
	• Government administration reflects Soviet type structure.	• Prepare and implement a program of reorganization of public administration in agriculture.
	• Research/education system has not been adjusted to emerging new conditions.	• Review and streamline agricultural education and research system.
	• Public activities (government research-education) in agriculture are seriously hampered by budgetary difficulties.	• Support the emergence of private farm advisory services.
	• Western type of agricultural extension system does not exist.	

UZBEKISTAN

Total Population	22 mil	Food and agriculture in GDP	27%	Agricultural output in 1996 as percentage	
Rural Population	60%	1996		of 1989 levels:	90%
		Food and agriculture in active	44%	Crop production	95%
Total Area	44.7mil ha	labor force (1996)	40%	Livestock production	80%
Agriculture area:	4.5 mil ha	Food and agriculture in export	20%	Share of livestock in agriculture (1996)	20%
Arable land	71%	(1996)		Agricultural area in private use (1996)	12%
Orchards	6.0%	in import (1996)		Share of family private farms in total	30%
Pasture & Forest	23.0%	Second largest exporter of cotton		agricultural output (1996)	
Irrigated	89%	lint.			

ISSUES	STATUS OF REFORMS	OBJECTIVES PROPOSED ACTIONS
1. Macro-economic Framework for Agriculture A. Production controls B. Prices/Subsidies	**Government committed to transformation to market economy, but slowly.** • Original policy was to phase out state order system by 1998. This was delayed and production targets for cotton and wheat remain in force at the district level. Production of other crops and livestock products has been liberalized. • Central price controls retained for part of cotton and wheat crops and *de facto* for other commodities at province/district levels. Subsidies on inputs eliminated but maintained on fuel and water.	**Distortion free, efficient and internationally competitive agricultural sector.** • Accelerate liberalization of production of all commodities. • Fully liberalize output markets and eliminate all subsidies on inputs.
C. Trade Policies	• Traditional international markets for cotton maintained. • Commodity "associations" maintain controlling influence over most marketed products. • Little trading outside FSU countries for non-cotton products. • Continuing traditional production practices cause major marketing problems and high crop losses at farm level.	• Remove all quantitative (non-tariff) trade restrictions • Permit emergence of private sector processors and traders by privatizing state owned agribusinesses and encouraging new entrants.
D. Taxation	• Government policy is to phase out taxation on primary agricultural products. • Net transfer from cotton and wheat revenues to the economy is 30%.	• Develop a transparent agricultural taxation based on business profits without discrimination.

ISSUE	STATUS OF REFORMS	OBJECTIVES PROPOSED ACTIONS
2. Land Reform and Farm Restructuring	**Government committed to transforming agriculture into an efficient and dynamic sector.** • Constitution restricts land holding rights to lease hold only; • Lack of coherent policy and absence of information system; • Slow progress in transforming collective farms to private operators;	**A farming system based on private ownership or long-term leases with irrevocable rights to inherit, trade or sub-lease.** • Provide information service for rural communities; • Make transparent evaluation and distribution of farmland and property and provide for trading of land and property rights; • Encourage voluntary grouping of "shareholders" for managing privatized farmland in blocks
3. Competitive Agroprocessing and Services for Agriculture	**Government intends to transfer processing enterprises to private control, improve efficiency of utilization and conserve water supplies, and liberalize the input sub-sector.** • Little progress due to uncertainty over procedures for decentralizing processing enterprises; • Proposing to revise water law and require creation of water users associations • Delayed introduction and enforcement of economic water charges • Production of agricultural chemicals remains under monopoly control; • Despite deregulation, inputs distribution remain effectively under monopoly control.	**Efficient, privately owned agrobusiness firms subject to market forces, and agroprocessing industries with high quality products which can compete in world markets. Full transition of production and distribution functions, with free access of new foreign and domestic operators. Enactment and enforcement of law to improve management and conservation of scarce water resources.** • Specify strategy for divesting cotton ginneries and other processing facilities; • Enable free entry of new private processing enterprises • Decentralized management of water resources and support creation of water users associations on all farms; • Apply water charges sufficient to ensure independent supply and maintenance of primary and secondary irrigation infrastructure • Introduce measures for disaggregating inputs monopoly to private enterprises;
4. Rural Financing	**Government policy is to reorganize financial services on commercial lines.** • Currently preparing plans for reorganization of state owned banks • Financial services in disarray; most banks insolvent and limited access by rural communities	**Viable financial institutions serving the agricultural and rural sector efficiently.** • Identify viable commercial banking operations for developing rural financial services; • Encourage self-help credit associations; • Apply economic interest rates for all credit funds.

ISSUE	STATUS OF REFORMS	OBJECTIVES PROPOSED ACTIONS
5. Institutional Framework	**Government policy is to simplify structure and reduce size of agriculture and water ministry.** • Agriculture and water resources reorganized under single Ministry of Agriculture and Water Management (MAWM); • Existing departments MAWM being reorganized to improve support for farmers	**Efficient and effective public sector administration and support for private agriculture.** • Focus public services on policy formulation, demand driven research, information and evaluation; • Encourage evolution of private sector support services

C. Statistical Annex: Food and Agriculture in Central and Eastern Europe and the CIS

Key to Tables:

CEE: the former socialist countries of Central-Eastern Europe, including the Baltic republics of the former Soviet Union, Albania, the member republics of ex-Yugoslavia and Albania;

CIS: all CIS member countries of the former Soviet Union;

CEFTA: the five CEFTA member countries in 1996 (Poland, Hungary, Czech Republic, Slovak Republic; and Slovenia);

EU-A5: countries in the EU-accession process - Estonia, Poland, Czech Republic, Hungary, and Slovenia;

Other CEE: the remaining countries of Central-Eastern Europe;

Russia: the Russian Federation

Euro CIS: Belarus, Moldova, and Ukraine;

Caucasus: Georgia, Armenia, Azerbaijan;

Central Asia: Tajikistan, Turkmenistan, Kazakhstan, the Kyrgyz Republic, and Uzbekistan.

The source for all tables in the Statistical Annex is FAOstat

Table 1: Population, Agricultural Area, and Arable Land of Transition Countries, Absolute and in Comparison to World

	Population, 1996		Agricultural Area, 1994		Arable Land, 1994	
	in Mil	% of World	in mil ha	% of World	in Mil ha	% of World
TOTAL	414	7.3	649	13.3	267	19.7
Total CEE	129	2.3	74	1.5	50	3.7
Total CIS	286	5.0	575	11.8	217	16.0
CEFTA	66	1.2	32	0.7	24	1.8
EU-A5	39	0.7	28	0.6	19	1.4
Other CEE	24	0.4	13	0.3	7	0.5
Russia	148	2.6	220	4.5	130	9.6
Euro CIS	66	1.2	54	1.1	41	3.0
Caucasus	17	0.3	8	0.2	3	0.2
Central Asia	54	1.0	293	6.0	43	3.1

Table 2: Population, Agricultural Area, and Arable Land of Transition Countries, Absolute and in Comparison to Total Transition Countries

	Population, 1996		Agricultural Area, 1994		Arable Land, 1994	
	in Mil	% of TC	in mil ha	% of TC	in Mil ha	% of TC
TOTAL	414	100.0	649	100.0	267	100.0
Total CEE	129	31.1	74	11.4	50	18.6
Total CIS	284	68.9	575	88.6	217	81.4
CEFTA	66	16.0	32	4.9	24	9.0
EU-A5	39	9.4	28	4.3	19	7.2
Other CEE	24	5.8	13	2.0	7	2.5
Russia	148	35.8	220	33.9	130	48.9
Euro CIS	66	16.1	54	8.3	41	15.4
Caucasus	17	4.0	8	1.2	3	1.1
Central Asia	54	13.0	293	45.1	43	16.0

Table 3: GDP Development

	GDP 1995		average annual GDP growth rate		
	US$ bil	% of TC	1995*	1990-1995*	1997 forecast** (EBRD)
TOTAL	838	100	-1.5	-7.0	2.0
Total CEE	329	39	5.2	-0.7	3.5
Total CIS	509	61	-5.8	-10.6	1.0
CEFTA	242	29	5.6	0.3	4.5
EU-A5	65	8	5.1	-4.5	-1.1
Other CEE	22	3	1.9	n.a.	5.5
Russia	345	41	-4.0	-9.8	1.5
Euro CIS	104	12	-12.0	-13.3	-1.5
Caucasus	8	1	-7.2	-22.4	7.0
Central Asia	52	6	-5.7	-9.1	2.3

* weighted by 1995 GDP

** weighted by 1995 GDP, Albania 1997 growth rate assumed to be equal than in 1995

Table 4: Development of Agricultural GDP

	Agricultural GDP		Agricultural Population	
	change in %, 1990-1995*	in % of total GDP, 1995**	in Mil, 1996	in % of total Population, 1996
TOTAL	-5.9	10.8	74.0	17.8
Total CEE	-1.9	9.2	23.1	17.9
Total CIS	-7.5	11.8	50.8	17.8
CEFTA	-3.0	6.7	11.9	18.0
EU-A5	-1.7	15.8	5.7	14.8
Other CEE	7.6	17.4	5.5	22.8
Russia	-6.3	9.2	18.0	12.1
Euro CIS	-10.0	13.3	12.4	18.7
Caucasus	-17.3	36.7	4.2	25.0
Central Asia	-6.2	22.7	16.3	29.9

*Azerbajan, Croatia, Lithuania, Macedonia FYR, Moldova, Czech Republic, Yugoslav FR, Slovenia, Tajikistan, Turkmenistan are not available and not included

** Moldova is not available and not included

Table 5: Production Change (95/96 in % of 89/90)

	TOTAL	CEE	CIS
Cereals	69	89	59
Coarse Grain	70	98	55
Wheat	68	78	64
Maize	87	109	47
Oilcrops	92	113	82
Rape and Sunflower	101	117	92
Sugarbeet	61	80	53
Fruits	77	84	72
Vegetable	91	105	83
Milk	71	86	66
Ruminant Meat	63	76	60
Pork	67	91	47
Poultry	54	86	38

Table 6: Cereals

Table 6a: Area of Cereals Cultivation (million ha.)

	1989	1990	1991	1992	1993	1994	1995	1996
TOTAL	131.2	129.9	129.5	128.9	129.9	122.7	119.8	114.4
Total CEE	26.4	26.0	26.7	27.6	28.7	28.6	27.9	27.0
Total CIS	104.8	103.9	102.8	101.2	101.2	94.0	91.9	87.4
CEFTA	13.7	13.7	14.0	13.6	13.8	14.1	13.9	14.0
EU-A5	8.2	7.8	8.3	10.3	10.9	10.6	10.2	9.3
Other CEE	4.5	4.5	4.4	3.8	4.0	4.0	3.8	3.6
Russia				59.7	58.9	54.3	52.9	51.4
Euro CIS				15.8	16.4	15.4	16.3	14.7
Caucasus				1.1	1.2	1.0	1.0	1.2
Central Asia				24.7	24.7	23.3	21.7	20.1

Table 6b: Yields of Cereals (tons per ha.)

	1989	1990	1991	1992	1993	1994	1995	1996
TOTAL	2.19	2.33	2.00	2.02	1.99	1.86	1.75	1.72
Total CEE	3.77	3.60	3.88	2.72	2.80	3.01	3.27	3.00
Total CIS	1.79	2.01	1.52	1.83	1.76	1.51	1.29	1.33
CEFTA	3.98	3.87	3.96	2.99	3.04	3.17	3.44	3.37
EU-A5	3.41	3.26	3.41	2.22	2.36	2.66	2.88	2.32
Other CEE	3.77	3.36	4.52	3.08	3.14	3.36	3.70	3.35
Russia				1.74	1.63	1.45	1.17	1.32
Euro CIS				2.82	3.25	2.62	2.47	2.06
Caucasus				1.90	1.60	1.65	1.60	1.65
Central Asia				1.39	1.07	0.92	0.69	0.79

Table 6c: Total Production of Cereals (million tons)

	1989	1990	1991	1992	1993	1994	1995	1996
TOTAL	**288**	**303**	**259**	**260**	**258**	**228**	**210**	**197**
Total CEE	**100**	**94**	**104**	**75**	**80**	**86**	**91**	**81**
Total CIS	188	209	156	185	178	142	119	116
CEFTA	54	53	56	41	42	45	48	47
EU-A5	28	25	28	23	26	28	29	22
Other CEE	17	15	20	12	12	13	14	12
Russia				104	96	79	62	68
Euro CIS				45	53	40	40	30
Caucasus				2	2	2	2	2
Central Asia				34	26	21	15	16

Table 6d: Net Exports of Cereals (million tons)

	1989	1990	1991	1992	1993	1994	1995
TOTAL	**-39.2**	**-33.7**	**-36.9**	**-38.0**	**-27.3**	**-7.4**	**3.1**
Total CEE	**-1.9**	**-2.2**	**0.3**	**3.7**	**-7.2**	**-1.5**	**4.8**
Total CIS	-37.2	-31.5	-37.2	-41.7	-20.1	-5.8	-1.7
CEFTA	-1.9	-0.2	1.9	4.3	-3.7	-0.3	3.7
EU-A5	-1.0	-0.9	-2.2	-1.8	-2.9	-0.5	1.2
Other CEE	0.9	-1.1	0.5	1.2	-0.6	-0.7	-0.2
Russia				-29.8	-13.4	-2.7	-1.4
Euro CIS				-5.7	-4.1	-0.9	-0.6
Caucasus				-1.3	-1.8	-1.7	-1.6
Central Asia				-4.8	-0.7	-0.5	1.9

Table 7: Wheat

Table 7a: Area of Wheat Cultivation (millions of ha.)

	1989	1990	1991	1992	1993	1994	1995	1996
TOTAL	57.5	58.0	55.7	55.4	56.1	52.4	55.4	55.6
Total CEE	9.8	9.9	9.8	8.6	10.3	10.4	10.4	9.1
Total CIS	47.7	48.2	45.9	46.7	45.8	42.0	45.0	46.5
CEFTA	4.7	4.7	4.8	4.4	4.7	4.8	4.8	4.9
EU-A5	3.5	3.4	3.4	3.0	4.1	4.1	4.1	3.2
Other CEE	1.7	1.7	1.7	1.2	1.5	1.5	1.5	1.1
Russia				24.3	24.7	22.2	23.9	25.0
Euro CIS				6.7	6.2	4.9	6.0	6.6
Caucasus				0.6	0.7	0.6	0.6	0.6
Central Asia				15.1	14.2	14.3	14.5	14.3

Table 7b: Yields of Wheat (tons per ha.)

	1989	1990	1991	1992	1993	1994	1995	1996
TOTAL	2.23	2.47	1.98	2.10	2.03	1.80	1.73	1.62
Total CEE	4.16	4.21	3.90	3.21	3.06	3.30	3.49	3.09
Total CIS	1.83	2.11	1.57	1.90	1.79	1.43	1.32	1.32
CEFTA	4.57	4.63	4.48	3.66	3.47	3.90	3.98	3.81
EU-A5	3.84	3.68	2.97	2.61	2.48	2.59	2.96	1.96
Other CEE	3.68	4.11	4.08	3.10	3.34	3.35	3.35	3.14
Russia				1.90	1.77	1.45	1.26	1.40
Euro CIS				3.10	3.79	3.00	2.99	2.25
Caucasus				1.98	1.62	1.58	1.44	1.70
Central Asia				1.35	0.98	0.86	0.73	0.76

Table 7c: Production of Wheat (millions of tons)

	1989	1990	1991	1992	1993	1994	1995	1996
TOTAL	**128**	**143**	**110**	**116**	**114**	**94**	**96**	**90**
Total CEE	**41**	**41**	**38**	**28**	**31**	**34**	**36**	**28**
Total CIS	87	102	72	89	82	60	60	62
CEFTA	21	22	21	16	16	19	19	19
EU-A5	13	13	10	8	10	11	12	6
Other CEE	6	7	7	4	5	5	5	3
Russia				46	44	32	30	35
Euro CIS				21	24	15	18	15
Caucasus				1	1	1	1	1
Central Asia				20	14	12	11	11

Table 7d: Net Exports of Wheat (million of tons)

Year	1989	1990	1991	1992	1993	1994	1995	1996
TOTAL	**-13**	**-14**	**-18**	**-26**	**-14**	**-5**	**2**	
Total CEE	**0**	**1**	**2**	**0**	**-3**	**0**	**5**	
Total CIS	**-13**	**-14**	**-19**	**-25**	**-11**	**-5**	**-3**	
CEFTA	-1	0	1	1	-1	0	4	
EU-A5	0	0	-1	-1	-1	0	1	
Other CEE	1	0	1	0	-1	-1	0	
Russia				-17	-6	-1	-2	
Euro CIS				-3	-3	-1	-1	
Caucasus				-1	-2	-2	-2	
Central Asia				-4	0	-1	1	

Table 8: Course Grains

Table 8a: Area of Course Grains under Cultivation (million ha.)

	1989	1990	1991	1992	1993	1994	1995	1996
TOTAL	**73.0**	**71.2**	**73.1**	**72.8**	**73.1**	**69.7**	**63.9**	**58.3**
Total CEE	**16.5**	**16.1**	**16.8**	**19.0**	**18.4**	**18.2**	**17.5**	**17.9**
Total CIS	56.4	55.1	56.4	53.9	54.7	51.5	46.4	40.4
CEFTA	9.0	9.0	9.2	9.1	9.1	9.3	9.0	9.1
EU-A5	4.7	4.3	4.9	7.2	6.8	6.4	6.2	6.2
Other CEE	2.8	2.8	2.7	2.6	2.5	2.5	2.3	2.6
Russia				35.1	34.0	31.9	28.9	26.3
Euro CIS				9.1	10.2	10.5	10.2	8.1
Caucasus				0.4	0.5	0.4	0.4	0.6
Central Asia				9.2	10.1	8.7	6.9	5.5

Table 8b: Yields of Course Grains (tons per ha.)

	1989	1990	1991	1992	1993	1994	1995	1996
TOTAL	**2.15**	**2.21**	**2.01**	**1.94**	**1.95**	**1.90**	**1.77**	**1.82**
Total CEE	**3.54**	**3.23**	**3.88**	**2.49**	**2.65**	**2.84**	**3.14**	**2.96**
Total CIS	**1.75**	**1.91**	**1.45**	**1.75**	**1.71**	**1.56**	**1.25**	**1.32**
CEFTA	3.68	3.47	3.70	2.67	2.82	2.79	3.15	3.13
EU-A5	3.11	2.94	3.72	2.05	2.29	2.70	2.82	2.51
Other CEE	3.82	2.91	4.81	3.06	3.02	3.36	3.92	3.43
Russia				1.62	1.53	1.44	1.09	1.24
Euro CIS				2.62	2.91	2.44	2.16	1.90
Caucasus				1.78	1.56	1.76	1.80	1.60
Central Asia				1.40	1.13	0.96	0.55	0.78

Table 8c: Production of Course Grains (million tons)

	1989	1990	1991	1992	1993	1994	1995	1996
TOTAL	**157**	**157**	**147**	**142**	**143**	**132**	**113**	**106**
Total CEE	**59**	**52**	**65**	**47**	**49**	**52**	**55**	**53**
Total CIS	**99**	**105**	**82**	**94**	**94**	**81**	**58**	**53**
CEFTA	33	31	34	24	26	26	28	28
EU-A5	15	13	18	15	16	17	17	15
Other CEE	11	8	13	8	7	8	9	9
Russia				57	52	46	31	33
Euro CIS				24	30	25	22	15
Caucasus				1	1	1	1	1
Central Asia				13	11	8	4	4

Table 9: Maize

Table 9a: Area of Maize Cultivation (million ha.)

	1989	1990	1991	1992	1993	1994	1995	1996
TOTAL	**11.1**	**9.3**	**9.7**	**10.3**	**10.1**	**9.0**	**9.3**	**9.4**
Total CEE	**7.0**	**6.5**	**6.7**	**7.7**	**7.1**	**7.0**	**6.8**	**7.3**
Total CIS	4.1	2.8	3.0	2.6	2.9	1.9	2.5	2.1
CEFTA	1.3	1.3	1.4	1.5	1.5	1.5	1.3	1.4
EU-A5	3.3	2.9	3.1	4.0	3.6	3.5	3.5	3.8
Other CEE	2.3	2.3	2.2	2.2	2.1	2.1	2.0	2.2
Russia				0.8	0.8	0.5	0.6	0.8
Euro CIS				1.4	1.7	0.9	1.5	0.8
Caucasus				0.1	0.1	0.1	0.2	0.2
Central Asia				0.3	0.3	0.3	0.2	0.2

Table 9b: Yields of Maize (tons per ha.)

	1989	1990	1991	1992	1993	1994	1995	1996
TOTAL	**3.81**	**3.24**	**4.50**	**2.76**	**3.00**	**3.15**	**3.54**	**3.18**
Total CEE	**3.86**	**3.13**	**5.03**	**2.78**	**2.96**	**3.41**	**3.82**	**3.45**
Total CIS	**3.71**	**3.48**	**3.29**	**2.68**	**3.11**	**2.19**	**2.79**	**2.26**
CEFTA	6.12	4.12	6.43	3.72	3.73	3.95	4.56	4.79
EU-A5	2.74	2.78	4.23	2.17	2.50	3.08	3.31	2.86
Other CEE	4.15	3.03	5.29	3.24	3.21	3.58	4.25	3.65
Russia				2.64	3.05	1.70	2.70	1.38
Euro CIS				2.50	3.09	2.31	2.93	3.19
Caucasus				2.23	2.12	2.42	2.59	1.75
Central Asia				3.80	3.71	2.49	2.30	2.48

Table 9c: Production of Maize (million tons)

	1989	1990	1991	1992	1993	1994	1995	1996
TOTAL	42.3	30.1	43.7	28.4	30.2	28.3	33.0	30.0
Total CEE	27.0	20.2	33.9	21.3	21.1	24.0	26.0	25.3
Total CIS	15.3	9.9	9.8	7.1	9.1	4.2	7.1	4.8
CEFTA	8.2	5.3	8.9	5.6	5.4	5.9	5.9	6.5
EU-A5	9.0	8.0	13.3	8.6	9.0	10.7	11.7	10.8
Other CEE	9.7	7.0	11.7	7.1	6.7	7.4	8.3	8.0
Russia				2.1	2.4	0.9	1.7	1.1
Euro CIS				3.5	5.2	2.2	4.4	2.6
Caucasus				0.2	0.3	0.4	0.4	0.4
Central Asia				1.2	1.2	0.8	0.5	0.6

Table 9d: Net Exports of Maize (Million tons)

	1989	1990	1991	1992	1993	1994	1995
TOTAL	-12.7	-13.5	-17.9	-25.6	-14.4	-5.0	1.7
Total CEE	0.4	0.8	1.6	-0.2	-3.2	-0.3	5.2
Total CIS	-13.1	-14.3	-19.4	-25.4	-11.2	-4.7	-3.5
CEFTA	-0.8	0.3	1.4	1.3	-1.1	0.3	4.1
EU-A5	0.4	0.1	-0.5	-1.3	-1.4	0.0	1.1
Other CEE	0.8	0.4	0.7	-0.1	-0.7	-0.6	0.0
Russia				-17.1	-6.3	-1.3	-2.1
Euro CIS				-3.3	-3.0	-0.5	-0.5
Caucasus				-1.3	-1.7	-1.8	-1.6
Central Asia				-3.7	-0.1	-1.1	0.8

Table 10: Oil Crops

Table 10a: Area of Oil Crops Cultivation (millions ha.)

	1989	1990	1991	1992	1993	1994	1995	1996
TOTAL	**18.9**	**18.2**	**17.1**	**9.9**	**9.2**	**9.4**	**11.5**	**11.2**
Total CEE	**3.2**	**2.7**	**2.6**	**3.0**	**2.7**	**2.8**	**3.4**	**3.3**
Total CIS	15.7	15.5	14.5	6.9	6.5	6.6	8.1	7.9
CEFTA	1.3	1.3	1.3	1.2	1.1	1.2	1.6	1.3
EU-A5	1.4	1.0	1.0	1.4	1.2	1.2	1.5	1.6
Other CEE	0.4	0.4	0.4	0.4	0.4	0.3	0.3	0.4
Russia				4.2	4.1	4.1	5.2	5.0
Euro CIS				2.3	2.1	2.2	2.5	2.5
Caucasus				0.0	0.0	0.0	0.0	0.1
Central Asia				0.3	0.3	0.3	0.4	0.4

Table 10b: Yields of Oil Crops (tons per ha.)

	1989	1990	1991	1992	1993	1994	1995	1996
TOTAL	**321**	**302**	**296**	**494**	**475**	**443**	**510**	**420**
Total CEE	**596**	**597**	**650**	**538**	**522**	**567**	**612**	**559**
Total CIS	**266**	**250**	**232**	**475**	**456**	**392**	**466**	**362**
CEFTA	821	749	763	664	645	664	721	673
EU-A5	382	428	487	429	398	478	494	439
Other CEE	604	550	689	533	571	539	589	652
Russia				339	314	284	351	270
Euro CIS				463	463	335	514	406
Caucasus				1917	1876	1648	755	574
Central Asia				2163	2145	1975	1582	1288

Table 10c: Production of Oil Crops (million tons)

	1989	1990	1991	1992	1993	1994	1995	1996
TOTAL	6071	5476	5071	4878	4393	4169	5883	4700
Total CEE	1894	1611	1720	1622	1415	1565	2104	1848
Total CIS	4177	3865	3351	3255	2979	2604	3779	2852
CEFTA	1095	943	986	805	698	803	1186	890
EU-A5	541	430	478	595	492	579	719	688
Other CEE	258	238	255	223	225	183	199	271
Russia				1432	1273	1174	1811	1340
Euro CIS				1053	993	725	1286	1008
Caucasus				42	30	33	29	31
Central Asia				729	682	672	652	473

Table 10d: Net Exports of Oil Crops (million tons)

	1989	1990	1991	1992	1993	1994	1995
TOTAL	-0.6	-0.5	-0.2	0.4	0.7	0.6	1.0
Total CEE	0.2	-0.2	0.3	0.2	0.3	0.0	0.6
Total CIS	-0.8	-0.3	-0.5	0.1	0.4	0.6	0.4
CEFTA	0.5	0.6	0.7	0.4	0.3	0.1	0.6
EU-A5	-0.1	-0.5	-0.3	0.0	0.0	-0.1	0.1
Other CEE	-0.3	-0.3	-0.1	-0.1	0.0	0.0	0.0
Russia				-0.1	0.1	0.2	0.3
Euro CIS				0.2	0.3	0.4	0.1
Caucasus				0.0	0.0	0.0	0.0
Central Asia				0.0	0.0	0.0	0.0

Table 11: Rape and Sunflower Seed

Table 11a: Area of Rape and Sunflower Seed Cultivation (millions ha.)

	1989	1990	1991	1992	1993	1994	1995	1996
TOTAL	**7.1**	**7.1**	**7.0**	**7.7**	**7.5**	**8.0**	**10.2**	**9.8**
Total CEE	**2.1**	**2.0**	**2.1**	**2.5**	**2.3**	**2.4**	**3.1**	**2.9**
Total CIS	5.0	5.1	4.9	5.2	5.2	5.6	7.1	6.9
CEFTA	1.1	1.1	1.1	1.1	1.0	1.1	1.5	1.2
EU-A5	0.7	0.7	0.8	1.1	1.1	1.1	1.3	1.4
Other CEE	0.3	0.3	0.2	0.3	0.3	0.2	0.2	0.3
Russia				3.0	3.0	3.3	4.4	4.2
Euro CIS				1.8	1.8	1.9	2.3	2.3
Caucasus				0.0	0.0	0.0	0.0	0.1
Central Asia				0.3	0.3	0.3	0.4	0.4

Table 11b: Yields of Rape and Sunflower Seed (tons per ha.)

	1989	1990	1991	1992	1993	1994	1995	1996
TOTAL	**1.69**	**1.51**	**1.43**	**1.29**	**1.18**	**1.06**	**1.27**	**1.04**
Total CEE	**2.14**	**1.92**	**1.91**	**1.60**	**1.47**	**1.60**	**1.72**	**1.55**
Total CIS	**1.51**	**1.35**	**1.21**	**1.15**	**1.05**	**0.82**	**1.07**	**0.82**
CEFTA	2.48	2.26	2.22	1.90	1.81	1.92	2.08	1.92
EU-A5	1.63	1.39	1.40	1.25	1.07	1.26	1.30	1.15
Other CEE	1.97	1.88	2.06	1.75	1.84	1.71	1.77	2.05
Russia				1.08	0.94	0.82	0.98	0.74
Euro CIS				1.37	1.33	0.91	1.39	1.09
Caucasus				0.58	0.23	0.49	0.22	0.15
Central Asia				0.57	0.44	0.37	0.28	0.20

Table 11c: Production of Rape and Sunflower (million tons)

	1989	1990	1991	1992	1993	1994	1995	1996
TOTAL	**11974**	**10813**	**9997**	**9942**	**8856**	**8429**	**12887**	**10162**
Total CEE	**4480**	**3905**	**4103**	**3956**	**3455**	**3867**	**5285**	**4500**
Total CIS	**7494**	**6909**	**5894**	**5987**	**5400**	**4563**	**7602**	**5662**
CEFTA	2840	2442	2542	2104	1829	2115	3179	2316
EU-A5	1132	956	1055	1381	1142	1377	1714	1637
Other CEE	508	507	505	470	484	375	392	547
Russia				3274	2862	2676	4323	3116
Euro CIS				2524	2407	1755	3157	2466
Caucasus				7	3	9	8	8
Central Asia				181	129	123	113	72

Table 12: Sugarbeet

Table 12a: Area of Sugarbeet Cultivation (millions ha.)

	1989	1990	1991	1992	1993	1994	1995	1996
TOTAL	**4.52**	**4.39**	**4.24**	**4.18**	**3.94**	**3.68**	**3.64**	**3.44**
Total CEE	**1.17**	**1.11**	**1.07**	**1.03**	**0.87**	**0.90**	**0.90**	**1.00**
Total CIS	3.34	3.29	3.17	3.15	3.07	2.78	2.74	2.44
CEFTA	0.73	0.74	0.69	0.66	0.64	0.63	0.64	0.72
EU-A5	0.30	0.20	0.24	0.25	0.15	0.18	0.18	0.18
Other CEE	0.15	0.16	0.15	0.12	0.07	0.09	0.09	0.10
Russia				1.44	1.33	1.10	1.09	1.02
Euro CIS				1.62	1.66	1.61	1.59	1.38
Caucasus				0.00	0.00	0.00	0.00	0.00
Central Asia				0.09	0.08	0.07	0.05	0.04

Table 12b: Yields of Sugarbeets (tons per ha.)

	1989	1990	1991	1992	1993	1994	1995	1996
TOTAL	**30.6**	**27.3**	**23.8**	**20.6**	**23.5**	**19.5**	**22.4**	**22.0**
Total CEE	**34.9**	**33.5**	**32.4**	**26.5**	**32.8**	**29.6**	**32.2**	**33.3**
Total CIS	**29.1**	**25.3**	**20.8**	**18.7**	**20.8**	**16.3**	**19.2**	**17.4**
CEFTA	35.9	36.5	33.2	29.7	36.7	30.9	35.3	36.0
EU-A5	26.1	19.4	23.2	16.8	19.6	23.1	21.9	21.6
Other CEE	47.5	37.4	43.7	30.1	26.8	32.5	29.9	35.7
Russia				17.8	19.1	12.6	17.6	15.9
Euro CIS				19.7	22.6	19.1	20.6	18.6
Caucasus				30.0	19.9	8.2	15.6	8.0
Central Asia				15.6	14.0	8.2	8.8	11.9

Table 12c: Production of Sugar Beets (million tons)

	1989	1990	1991	1992	1993	1994	1995	1996
TOTAL	**138.3**	**120.1**	**100.7**	**86.1**	**92.4**	**71.9**	**81.6**	**75.7**
Total CEE	**40.9**	**37.1**	**34.8**	**27.3**	**28.4**	**26.7**	**29.1**	**33.4**
Total CIS	**97.4**	**83.0**	**65.9**	**58.9**	**64.0**	**45.2**	**52.5**	**42.3**
CEFTA	26.1	27.1	22.8	19.5	23.4	19.6	22.7	26.0
EU-A5	7.7	3.9	5.6	4.3	3.0	4.1	3.9	4.0
Other CEE	7.1	6.2	6.4	3.5	2.0	3.0	2.5	3.4
Russia				25.5	25.5	13.9	19.1	16.1
Euro CIS				31.9	37.4	30.7	32.9	25.6
Caucasus				0.0	0.0	0.0	0.0	0.0
Central Asia				1.4	1.1	0.5	0.5	0.5

Table 13: Milk

Table 13a: Number of Cows, 1989-1996 (Million Head)

Year	1989	1990	1991	1992	1993	1994	1995	1996
TOTAL	62.7	62.1	60.8	54.6	51.6	50.7	48.2	47.7
Total CEE	17.5	17.1	16.2	15.2	13.6	13.0	12.3	12.2
Total CIS	45.2	45.0	44.6	39.3	38.0	37.7	35.8	35.5
CEFTA	9.4	9.3	8.8	8.4	7.3	6.8	6.5	6.3
EU-A5	1.6	1.5	1.5	3.1	2.9	2.6	2.5	2.5
Other CEE	6.5	6.2	5.9	3.8	3.4	3.6	3.4	3.4
Russia				20.2	19.9	19.4	18.1	18.1
Euro CIS				11.0	11.6	10.6	10.3	9.6
Caucasus				1.5	1.5	1.5	1.5	1.5
Central Asia				6.7	6.1	6.3	6.0	6.3

Table 13b: Milk Yields 1989 -1996 (Liters per Cow)

Year	1989	1990	1991	1992	1993	1994	1995	1996
TOTAL	2341	2346	2224	2232	2278	2210	2223	2104
Total CEE	2169	2170	2096	2343	2451	2515	2643	2647
Total CIS	2407	2413	2270	2189	2216	2105	2078	1918
CEFTA	2806	2758	2590	2536	2735	2803	2880	2920
EU-A5	3964	4049	4136	3451	3393	3773	4014	3972
Other CEE	803	826	838	1017	1051	1059	1188	1207
Russia				2339	2340	2174	2173	1975
Euro CIS				2386	2347	2318	2261	2260
Caucasus				1197	1087	1094	1122	1187
Central Asia				1635	1849	1773	1719	1413

Table 13c: Milk Production (Million Tons)

Year	1989	1990	1991	1992	1993	1994	1995	1996
TOTAL	146.8	145.7	135.2	121.8	117.6	112.0	107.0	100.3
Total CEE	38.0	37.2	33.9	35.7	33.3	32.6	32.5	32.2
Total CIS	108.8	108.5	101.3	86.1	84.3	79.4	74.5	68.1
CEFTA	26.5	25.7	22.9	21.3	20.1	19.1	18.7	18.3
EU-A5	6.3	6.3	6.1	10.5	9.7	9.7	9.8	9.8
Other CEE	5.2	5.2	4.9	3.8	3.6	3.8	4.0	4.2
Russia				47.2	46.5	42.2	39.3	35.7
Euro CIS				26.1	24.9	24.6	23.2	21.6
Caucasus				1.7	1.6	1.6	1.7	1.8
Central Asia				11.0	11.2	11.1	10.3	9.0

Table 13d: Net Exports of Milk (Million Tons)

Year	1989	1990	1991	1992	1993	1994	1995	1996
TOTAL	-1	-1	-1	1	1	2	1	
Total CEE	1	1	2	3	2	3	2	
Total CIS	-2	-2	-2	-2	-2	-1	-1	
CEFTA	1	2	2	2	2	2	2	
EU-A5	0	0	0	1	1	1	1	
other CEE	0	0	0	0	0	0	0	
Russia				-2	-2	-1	-2	
Euro CIS				0	0	1	1	
Caucasus				0	0	0	0	
Central Asia				0	0	0	0	

Table 14: Ruminant Meat

Table 14a: Total Number of Cattle (Million Head)

Year	1989	1990	1991	1992	1993	1994	1995	1996
TOTAL	150.6	148.4	143.0	135.3	128.4	121.9	111.0	101.2
Total CEE	31.0	30.0	27.3	28.4	25.1	23.4	21.7	21.6
Total CIS	119.6	118.4	115.6	107.0	103.3	98.5	89.3	79.5
CEFTA	17.5	16.8	15.3	14.5	13.0	12.3	11.6	11.7
EU-A5	8.0	7.9	6.8	10.0	8.1	7.2	6.2	6.1
other CEE	5.5	5.3	5.2	3.9	3.9	3.8	3.8	3.8
Russia				54.7	52.2	48.9	43.3	39.7
Euro CIS				31.3	29.6	28.4	25.9	21.4
Caucasus				3.6	3.2	3.1	3.1	3.1
Central Asia				17.4	18.2	18.2	17.0	15.3

Table 14b: Total Number of Sheep (Million Head)

Year	1989	1990	1991	1992	1993	1994	1995	1996
TOTAL	182.3	178.7	170.2	161.3	150.7	138.9	115.2	98.4
Total CEE	41.6	40.1	37.3	33.8	29.0	26.5	24.9	24.2
Total CIS	140.7	138.6	132.9	127.5	121.7	112.5	90.2	74.2
CEFTA	7.7	7.3	6.1	4.6	3.7	2.7	2.3	2.1
EU-A5	24.8	23.6	22.0	21.0	17.2	15.5	14.5	13.9
other CEE	9.2	9.2	9.1	8.3	8.0	8.2	8.2	8.2
Russia				52.2	48.2	40.6	31.8	25.8
Euro CIS				8.9	8.2	7.8	6.4	3.7
Caucasus				7.5	6.7	6.0	5.8	5.6
Central Asia				59.0	58.6	58.0	46.2	39.1

Table 14c: Production of Ruminant Meat (Million Tons)

Year	1989	1990	1991	1992	1993	1994	1995	1996
TOTAL	11.9	12.1	11.3	10.1	9.3	9.1	7.7	7.4
Total CEE	2.1	2.3	2.1	2.3	2.1	1.8	1.7	1.7
Total CIS	9.8	9.8	9.1	7.7	7.2	7.3	6.0	5.7
CEFTA	1.2	1.3	1.2	1.1	0.9	0.8	0.7	0.7
EU-A5	0.5	0.6	0.6	0.9	0.8	0.7	0.6	0.5
Other CEE	0.4	0.4	0.4	0.4	0.4	0.4	0.4	0.4
Russia				4.0	3.7	3.6	3.0	2.7
Euro CIS				2.3	1.9	1.9	1.6	1.4
Caucasus				0.2	0.1	0.2	0.2	0.2
Central Asia				1.3	1.5	1.7	1.3	1.4

Table 14d: Net Exports of Ruminant Meat (Million Tons)

Year	1989	1990	1991	1992	1993	1994	1995
TOTAL	-0.5	-0.6	-0.7	-0.4	-0.5	-0.4	-0.4
Total CEE	0.0	0.0	0.1	0.1	0.1	0.0	0.0
Total CIS	-0.4	-0.6	-0.7	-0.5	-0.6	-0.3	-0.4
CEFTA	0.0	0.1	0.1	0.1	0.0	0.0	0.0
EU-A5	0.0	-0.1	0.0	0.1	0.1	0.0	0.0
Other CEE	0.0	-0.1	0.0	0.0	0.0	0.0	0.0
Russia				-0.7	-0.7	-0.6	-0.6
Euro CIS				0.3	0.2	0.3	0.3
Caucasus				0.0	0.0	0.0	0.0
Central Asia				-0.1	0.0	0.0	0.0

Table 15: Pork

Table 15a: Total Number of Pigs (Million Head)

	1989	1990	1991	1992	1993	1994	1995	1996
TOTAL	138.7	137.1	136.1	124.8	110.8	103.2	96.0	93.0
Total CEE	60.6	58.1	60.7	59.9	53.0	50.3	49.0	49.8
Total CIS	78.1	79.0	75.4	64.9	57.7	52.9	47.0	43.2
CEFTA	34.5	34.6	37.0	35.7	31.7	31.3	31.2	30.5
EU-A5	18.5	16.0	16.2	18.3	15.3	13.4	12.0	12.2
Other CEE	7.6	7.5	7.5	5.9	6.0	5.5	5.8	7.1
Russia				35.4	31.5	28.6	24.9	22.6
Euro CIS				24.3	22.0	20.6	19.0	18.1
Caucasus				1.1	0.6	0.5	0.5	0.5
Central Asia				4.1	3.6	3.2	2.6	2.0

Table 15b: Production of Pork (Million Tons)

Year	1989	1990	1991	1992	1993	1994	1995	1996
TOTAL	12.5	12.5	11.6	10.6	9.6	8.5	8.6	8.2
Total CEE	5.8	5.8	5.6	5.8	5.5	5.0	5.4	5.2
Total CIS	6.7	6.7	6.0	4.8	4.1	3.6	3.3	3.0
CEFTA	3.8	3.8	3.7	3.7	3.5	3.1	3.4	3.3
EU-A5	1.2	1.2	1.2	1.4	1.3	1.2	1.2	1.1
Other CEE	0.8	0.8	0.7	0.8	0.7	0.7	0.7	0.7
Russia				2.8	2.4	2.1	1.9	1.7
Euro CIS				1.6	1.4	1.2	1.2	1.1
Caucasus				0.1	0.1	0.1	0.1	0.1
Central Asia				0.3	0.3	0.2	0.2	0.1

Table 15c: Net Exports of Pork (Million Tons)

Year	1989	1990	1991	1992	1993	1994	1995	1996
TOTAL	-0.7	-1.4	-0.7	0.9	0.5	1.9	1.2	
Total CEE	1.1	1.1	1.6	2.5	2.4	2.6	2.3	
Total CIS	-1.8	-2.5	-2.3	-1.6	-1.8	-0.7	-1.1	
CEFTA	1.1	1.7	2.0	2.2	2.2	1.9	1.8	
EU-A5	0.1	-0.2	0.0	0.6	0.5	1.1	0.9	
Other CEE	-0.1	-0.4	-0.3	-0.3	-0.3	-0.3	-0.4	
Russia				-1.7	-1.8	-0.9	-1.9	
Euro CIS				0.3	0.3	0.6	1.0	
Caucasus				-0.1	-0.2	-0.2	-0.2	
Central Asia				-0.2	-0.2	-0.2	-0.1	

Table 16: Poultry

Table 16a: Total Number of Poultry (Million Head)

Year	1989	1990	1991	1992	1993	1994	1995	1996
TOTAL	1.64	1.59	1.56	1.52	1.32	1.23	1.08	1.01
Total CEE	0.44	0.42	0.40	0.38	0.33	0.30	0.31	0.31
Total CIS	1.20	1.18	1.16	1.13	0.99	0.92	0.78	0.70
CEFTA	0.18	0.18	0.16	0.16	0.15	0.14	0.14	0.14
EU-A5	0.18	0.16	0.16	0.17	0.13	0.12	0.11	0.12
Other CEE	0.08	0.08	0.08	0.05	0.05	0.05	0.05	0.05
Russia				0.65	0.57	0.57	0.47	0.42
Euro CIS				0.31	0.27	0.24	0.22	0.20
Caucasus				0.05	0.04	0.03	0.03	0.03
Central Asia				0.12	0.10	0.08	0.06	0.04

Table 16b: Production of Poultry Meat (Million tons)

Year	1989	1990	1991	1992	1993	1994	1995	1996
TOTAL	4.94	4.93	4.48	3.68	3.27	2.89	2.72	2.63
Total CEE	1.64	1.65	1.39	1.32	1.27	1.28	1.41	1.41
Total CIS	3.30	3.28	3.09	2.36	2.00	1.61	1.31	1.22
CEFTA	0.85	0.81	0.74	0.74	0.71	0.77	0.85	0.86
EU-A5	0.53	0.57	0.43	0.45	0.44	0.38	0.43	0.42
Other CEE	0.26	0.27	0.22	0.13	0.12	0.12	0.13	0.13
Russia				1.43	1.28	1.07	0.86	0.80
Euro CIS				0.66	0.52	0.38	0.33	0.31
Caucasus				0.06	0.04	0.04	0.04	0.05
Central Asia				0.21	0.16	0.12	0.08	0.07

Table 16c: Net Exports of Poultry Meat (Million tons)

Year	1989	1990	1991	1992	1993	1994	1995	1996
TOTAL	0.1	-0.1	0.0	0.0	-0.2	-0.6	-0.9	
Total CEE	0.3	0.2	0.1	0.1	0.0	0.0	0.0	
Total CIS	-0.1	-0.3	-0.2	-0.1	-0.2	-0.6	-0.9	
CEFTA	0.2	0.2	0.2	0.1	0.0	0.1	0.1	
EU-A5	0.0	0.0	0.0	0.0	0.0	0.0	0.0	
Other CEE	0.0	0.0	0.0	0.0	0.0	0.0	0.0	
Russia				-0.1	-0.2	-0.5	-0.9	
Euro CIS				0.0	0.0	0.0	0.0	
Caucasus				0.0	0.0	0.0	0.0	
Central Asia				0.0	0.0	0.0	0.0	

Distributors of World Bank Publications

Prices and credit terms vary from country to country. Consult your local distributor before placing an order.

ARGENTINA
Oficina del Libro Internacional
Av. Cordoba 1877
1120 Buenos Aires
Tel: (54 11) 815-8354
Fax: (54 11) 815-8156
E-mail: olilibro@satlink.com

AUSTRALIA, FIJI, PAPUA NEW GUINEA, SOLOMON ISLANDS, VANUATU, AND SAMOA
D.A. Information Services
648 Whitehorse Road
Mitcham 3132
Victoria
Tel: (61) 3 9210 7777
Fax: (61) 3 9210 7788
E-mail: service@dadirect.com.au

AUSTRIA
Gerold and Co.
Weihburggasse 26
A-1011 Wien
Tel: (43 1) 512-47-31-0
Fax: (43 1) 512-47-31-29

BANGLADESH
Micro Industries Development Assistance Society (MIDAS)
House 5, Road 16
Dhanmondi R/Area
Dhaka 1209
Tel: (880 2) 326427
Fax: (880 2) 811188

BELGIUM
Jean De Lannoy
Av. du Roi 202
1060 Brussels
Tel: (32 2) 538-5169
Fax: (32 2) 538-0841

BRAZIL
Publicacões Tecnicas Internacionais Ltda.
Rua Peixoto Gomide, 209
01409 Sao Paulo, SP.
Tel: (55 11) 259-6644
Fax: (55 11) 258-6990
E-mail: postmaster@pti.uol.br

CANADA
Renouf Publishing Co. Ltd.
5369 Canotek Road
Ottawa, Ontario K1J 9J3
Tel: (613) 745-2665
Fax: (613) 745-7660
E-mail: order.dept@renoufbooks.com

CHINA
China Financial & Economic Publishing House
8, Da Fo Si Dong Jie
Beijing
Tel: (86 10) 6333-8257
Fax: (86 10) 6401-7365

China Book Import Centre
P.O. Box 2825
Beijing

COLOMBIA
Infoenlace Ltda.
Carrera 6 No. 51-21
Apartado Aereo 34270
Santafé de Bogotá, D.C.
Tel: (57 1) 285-2798
Fax: (57 1) 285-2798

COTE D'IVOIRE
Center d'Edition et de Diffusion Africaines (CEDA)
04 B.P. 541
Abidjan 04
Tel: (225) 24 6510;24 6511
Fax: (225) 25 0567

CYPRUS
Center for Applied Research
Cyprus College
6, Diogenes Street, Engomi
P.O. Box 2006
Nicosia
Tel: (357 2) 44-1730
Fax: (357 2) 46-2051

CZECH REPUBLIC
USIS, NIS Prodejna
Havelkova 22
130 00 Prague 3
Tel: (420 2) 2423 1486
Fax: (420 2) 2423 1114

DENMARK
SamfundsLitteratur
Rosenoerns Allé 11
DK-1970 Frederiksberg C
Tel: (45 31) 351942
Fax: (45 31) 357822

ECUADOR
Libri Mundi
Libreria Internacional
P.O. Box 17-01-3029
Juan Leon Mera 851
Quito
Tel: (593 2) 521-606; (593 2) 544-185
Fax: (593 2) 504-209
E-mail: librimu1@librimundi.com.ec

CODEU
Ruiz de Castilla 763, Edif. Expocolor
Primer piso, Of. #2
Quito
Tel/Fax: (593 2) 507-383; 253-091
E-mail: codeu@impsat.net.ec

EGYPT, ARAB REPUBLIC OF
Al Ahram Distribution Agency
Al Galaa Street
Cairo
Tel: (20 2) 578-6083
Fax: (20 2) 578-6833

The Middle East Observer
41, Sherif Street
Cairo
Tel: (20 2) 393-9732
Fax: (20 2) 393-9732

FINLAND
Akateeminen Kirjakauppa
P.O. Box 128
FIN-00101 Helsinki
Tel: (358 0) 121 4418
Fax: (358 0) 121-4435
E-mail: akatilaus@stockmann.fi

FRANCE
World Bank Publications
66, avenue d'Iéna
75116 Paris
Tel: (33 1) 40-69-30-56/57
Fax: (33 1) 40-69-30-68

GERMANY
UNO-Verlag
Poppelsdorfer Allee 55
53115 Bonn
Tel: (49 228) 949020
Fax: (49 228) 217492
E-mail: unoverlag@aol.com

GHANA
Epp Books Services
P.O. Box 44
TUC
Accra

GREECE
Papasotiriou S.A.
35, Stournara Str.
106 82 Athens
Tel: (30 1) 364-1826
Fax: (30 1) 364-8254

HAITI
Culture Diffusion
5, Rue Capois
C.P. 257
Port-au-Prince
Tel: (509) 23 9260
Fax: (509) 23 4858

HONG KONG, CHINA; MACAO
Asia 2000 Ltd.
Sales & Circulation Department
Seabird House, unit 1101-02
22-28 Wyndham Street, Central
Hong Kong
Tel: (852) 2530-1409
Fax: (852) 2526-1107
E-mail: sales@asia2000.com.hk

HUNGARY
Euro Info Service
Margitszgeti Europa Haz
H-1138 Budapest
Tel: (36 1) 350 80 24, 350 80 25
Fax: (36 1) 350 90 32
E-mail: euroinfo@mail.matav.hu

INDIA
Allied Publishers Ltd.
751 Mount Road
Madras - 600 002
Tel: (91 44) 852-3938
Fax: (91 44) 852-0649

INDONESIA
Pt. Indira Limited
Jalan Borobudur 20
P.O. Box 181
Jakarta 10320
Tel: (62 21) 390-4290
Fax: (62 21) 390-4289

IRAN
Ketab Sara Co. Publishers
Khaled Eslamboli Ave., 6th Street
Delafrooz Alley No. 8
P.O. Box 15745-733
Tehran 15117
Tel: (98 21) 8717819; 8716104
Fax: (98 21) 8712479
E-mail: ketab-sara@neda.net.ir

Kowkab Publishers
P.O. Box 19575-511
Tehran
Tel: (98 21) 258-3723
Fax: (98 21) 258-3723

IRELAND
Government Supplies Agency
Oifig an tSoláthair
4-5 Harcourt Road
Dublin 2
Tel: (353 1) 661-3111
Fax: (353 1) 475-2670

ISRAEL
Yozmot Literature Ltd.
P.O. Box 56055
3 Yohanan Hasandlar Street
Tel Aviv 61560
Tel: (972 3) 5285-397
Fax: (972 3) 5285-397

R.O.Y. International
P.O. Box 13056
Tel Aviv 61130
Tel: (972 3) 5461423
Fax: (972 3) 5461442
E-mail: royil@netvision.net.il

Palestinian Authority/Middle East
Index Information Services
P.O.B. 19502 Jerusalem
Tel: (972 2) 6271219
Fax: (972 2) 6271634

ITALY
Licosa Commissionaria Sansoni SPA
Via Duca Di Calabria, 1/1
Casella Postale 552
50125 Firenze
Tel: (55) 645-415
Fax: (55) 641-257
E-mail: licosa@ftbcc.it

JAMAICA
Ian Randle Publishers Ltd.
206 Old Hope Road, Kingston 6
Tel: 876-927-2085
Fax: 876-977-0243
E-mail: irpl@colis.com

JAPAN
Eastern Book Service
3-13 Hongo 3-chome, Bunkyo-ku
Tokyo 113
Tel: (81 3) 3818-0861
Fax: (81 3) 3818-0864
E-mail: orders@svt-ebs.co.jp

KENYA
Africa Book Service (E.A.) Ltd.
Quaran House, Mfangano Street
P.O. Box 45245
Nairobi
Tel: (254 2) 223 641
Fax: (254 2) 330 272

KOREA, REPUBLIC OF
Daejon Trading Co. Ltd.
P.O. Box 34, Youida, 706 Seoun Bldg
44-6 Youido-Dong, Yeongchengpo-Ku
Seoul
Tel: (82 2) 785-1631/4
Fax: (82 2) 784-0315

LEBANON
Librairie du Liban
P.O. Box 11-9232
Beirut
Tel: (961 9) 217 944
Fax: (961 9) 217 434

MALAYSIA
University of Malaya Cooperative Bookshop, Limited
P.O. Box 1127
Jalan Pantai Baru
59700 Kuala Lumpur
Tel: (60 3) 755-5000
Fax: (60 3) 755-4424
E-mail: umkoop@tm.net.my

MEXICO
INFOTEC
Av. San Fernando No. 37
Col. Toriello Guerra
14050 Mexico, D.F.
Tel: (52 5) 624-2800
Fax: (52 5) 624-2822
E-mail: infotec@rtn.net.mx

Mundi-Prensa Mexico S.A. de C.V.
c/Rio Panuco, 141-Colonia Cuauhtemoc
06500 Mexico, D.F.
Tel: (52 5) 533-5658
Fax: (52 5) 514-6799

NEPAL
Everest Media International Services (P) Ltd.
GPO Box 5443
Kathmandu
Tel: (977 1) 472 152
Fax: (977 1) 224 431

NETHERLANDS
De Lindeboom/InOr-Publikaties
P.O. Box 202, 7480 AE Haaksbergen
Tel: (31 53) 574-0004
Fax: (31 53) 572-9296
E-mail: lindeboo@worldonline.nl

NEW ZEALAND
EBSCO NZ Ltd.
Private Mail Bag 99914
New Market
Auckland
Tel: (64 9) 524-8119
Fax: (64 9) 524-8067

NIGERIA
University Press Limited
Three Crowns Building Jericho
Private Mail Bag 5095
Ibadan
Tel: (234 22) 41-1356
Fax: (234 22) 41-2056

NORWAY
NIC Info A/S
Book Department, Postboks 6512 Etterstad
N-0606 Oslo
Tel: (47 22) 97-4500
Fax: (47 22) 97-4545

PAKISTAN
Mirza Book Agency
65, Shahrah-e-Quaid-e-Azam
Lahore 54000
Tel: (92 42) 735 3601
Fax: (92 42) 576 3714

Oxford University Press
5 Bangalore Town
Sharae Faisal
PO Box 13033
Karachi-75350
Tel: (92 21) 446307
Fax: (92 21) 4547640
E-mail: ouppak@TheOffice.net

Pak Book Corporation
Aziz Chambers 21, Queen's Road
Lahore
Tel: (92 42) 636 3222; 636 0885
Fax: (92 42) 636 2328
E-mail: pbc@brain.net.pk

PERU
Editorial Desarrollo SA
Apartado 3824, Lima 1
Tel: (51 14) 285380
Fax: (51 14) 286628

PHILIPPINES
International Booksource Center Inc.
1127-A Antipolo St. Barangay, Venezuela
Makati City
Tel: (63 2) 896 6501; 6505; 6507
Fax: (63 2) 896 1741

POLAND
International Publishing Service
Ul. Piekna 31/37
00-677 Warzawa
Tel: (48 2) 628-6089
Fax: (48 2) 621-7255
E-mail: books%ips@ikp.atm.com.pl

PORTUGAL
Livraria Portugal
Apartado 2681, Rua Do Carmo 70-74
1200 Lisbon
Tel: (1) 347-4982
Fax: (1) 347-0264

ROMANIA
Compani De Librarii Bucuresti S.A.
Str. Lipscani no. 26, sector 3
Bucharest
Tel: (40 1) 613 9645
Fax: (40 1) 312 4000

RUSSIAN FEDERATION
Isdatelstvo <Ves Mir>
9a, Kolpachniy Pereulok
Moscow 101831
Tel: (7 095) 917 87 49
Fax: (7 095) 917 92 59

SINGAPORE; TAIWAN, CHINA; MYANMAR; BRUNEI
Ashgate Publishing Asia Pacific Pte. Ltd.
41 Kallang Pudding Road #04-03
Golden Wheel Building
Singapore 349316
Tel: (65) 741-5166
Fax: (65) 742-9356
E-mail: ashgate@asianconnect.com

SLOVENIA
Gospodarski Vestnik Publishing Group
Dunajska cesta 5
1000 Ljubljana
Tel: (386 61) 133 83 47; 132 12 30
Fax: (386 61) 133 80 30
E-mail: repansekj@gvestnik.si

SOUTH AFRICA, BOTSWANA
For single titles:
Oxford University Press Southern Africa
Vasco Boulevard, Goodwood
PO Box 12119, N1 City 7463
Cape Town
Tel: (27 21) 595 4400
Fax: (27 21) 595 4430
E-mail: oxford@oup.co.za

For subscription orders:
International Subscription Service
P.O. Box 41095
Craighall
Johannesburg 2024
Tel: (27 11) 880-1448
Fax: (27 11) 880-6248
E-mail: iss@is.co.za

SPAIN
Mundi-Prensa Libros, S.A.
Castello 37
28001 Madrid
Tel: (34 1) 431-3399
Fax: (34 1) 575-3998
E-mail: libreria@mundiprensa.es

Mundi-Prensa Barcelona
Consell de Cent, 391
08009 Barcelona
Tel: (34 3) 488-3492
Fax: (34 3) 487-7659
E-mail: barcelona@mundiprensa.es

SRI LANKA, THE MALDIVES
Lake House Bookshop
100, Sir Chittampalam Gardiner Mawatha
Colombo 2
Tel: (94 1) 32105
Fax: (94 1) 432104
E-mail: LHL@sri.lanka.net

SWEDEN
Wennergren-Williams AB
P.O. Box 1305
S-171 25 Solna
Tel: (46 8) 705-97-50
Fax: (46 8) 27-00-71
E-mail: mail@wwi.se

SWITZERLAND
Librairie Payot Service Institutionnel
Côtes-de-Montbenon 30
1002 Lausanne
Tel: (41 21) 341-3229
Fax: (41 21) 341-3235

ADECO Van Diermen EditionsTechniques
Ch. de Lacuez 41
CH1807 Blonay
Tel: (41 21) 943 2673
Fax: (41 21) 943 3605

THAILAND
Central Books Distribution
306 Silom Road
Bangkok 10500
Tel: (66 2) 235-5400
Fax: (66 2) 237-8321

TRINIDAD & TOBAGO AND THE CARRIBBEAN
Systematics Studies Ltd.
St. Augustine Shopping Center
Eastern Main Road, St. Augustine
Trinidad & Tobago, West Indies
Tel: (868) 645-8466
Fax: (868) 645-8467
E-mail: tobe@trinidad.net

UGANDA
Gustro Ltd.
PO Box 9997, Madhvani Building
Plot 16/4 Jinja Rd.
Kampala
Tel: (256 41) 251 467
Fax: (256 41) 251 468
E-mail: gus@swiftuganda.com

UNITED KINGDOM
Microinfo Ltd.
P.O. Box 3, Alton, Hampshire GU34 2PG
England
Tel: (44 1420) 86848
Fax: (44 1420) 89889
E-mail: wbank@ukminfo.demon.co.uk

The Stationery Office
51 Nine Elms Lane
London SW8 5DR
Tel: (44 171) 873-8400
Fax: (44 171) 873-8242

VENEZUELA
Tecni-Ciencia Libros, S.A.
Centro Cuidad Comercial Tamanco
Nivel C2, Caracas
Tel: (58 2) 959 5547; 5035; 0016
Fax: (58 2) 959 5636

ZAMBIA
University Bookshop, University of Zambia
Great East Road Campus
P.O. Box 32379
Lusaka
Tel: (260 1) 252 576
Fax: (260 1) 253 952

ZIMBABWE
Academic and Baobab Books (Pvt.) Ltd.
4 Conald Road, Graniteside
P.O. Box 567
Harare
Tel: 263 4 755035
Fax: 263 4 781913